I0502888

NIST Technical Note 1806

Improving the Performance of a Roof Top Air-Conditioning Unit by Refrigerant Circuitry Optimization

David Yashar
Sunil Lee
Energy and Environment Division
Engineering Laboratory

July 2013

U.S. Department of Commerce
Penny Pritzker, Secretary

National Institute of Standards and Technology
Patrick D. Gallagher, Under Secretary of Commerce for Standards and Technology and Director

Certain commercial entities, equipment, or materials may be identified in this document in order to describe an experimental procedure or concept adequately. Such identification is not intended to imply recommendation or endorsement by the National Institute of Standards and Technology, nor is it intended to imply that the entities, materials, or equipment are necessarily the best available for the purpose.

IMPROVING THE PERFORMANCE OF A ROOF TOP AIR-CONDITIONING UNIT BY REFRIGERANT CIRCUITRY OPTIMIZATION

David A. Yashar and Sunil Lee

National Institute of Standards and Technology

Gaithersburg, MD 20899

Abstract

This study demonstrates the performance improvement of an air-to-air roof top unit (RTU) achieved by optimizing an evaporator's refrigerant circuitry using evolutionary algorithms. The subject of this study is a unit with a cooling capacity of 7.5 Tons (26.4 kW). The RTU employs two separate refrigerant cycles having separate compressors, condensers, and thermostatic expansion valves (TXV) but using a single evaporator slab in which two separate refrigerant circuits are implemented.

We modified the RTU by replacing the refrigerant-to-air condensers with water cooled brazed plate heat exchangers in order to facilitate testing. Performance tests were conducted in a conditioned environmental chamber in line with AHRI standard 340/360-2007; in order to accomplish this, we maintained the liquid line saturation pressure and subcooling from the manufacturer's test data by adjusting the condenser water flow rate and temperature. We also measured the in-situ air velocity profile using Particle Image Velocimetry (PIV), a non-intrusive, laser-based technique. The measurements showed that the range of air velocities passing through the heat exchanger varied from 0.5 ms^{-1} to 3.0 ms^{-1}, with the area weighted average of the measurements being 1.75 ms^{-1}. The PIV data was used to generate a map of the air flow distribution through the heat exchanger, which served as the basis for refrigerant circuitry optimization.

We simulated the performance of the original evaporator using the measured air velocity distribution and The National Institute of Standards and Technology's (NIST) heat exchanger model, EVAP-COND, and tuned our computational model to exactly match the laboratory measurements. We then used the measured air velocity distribution with NIST's evolutionary algorithm optimization module, Intelligent System for Heat Exchanger Design (ISHED), to redesign the evaporator circuitry. The optimization process resulted in a design with a simulated capacity nearly 8 % higher than the original design.

The RTU manufacturer produced a new evaporator implementing the optimized refrigerant circuitry. We replaced the original evaporator with the prototype of the optimized evaporator and measured the system's performance. The system with the optimized evaporator showed an improvement of 2.2 % in capacity and 2.9 % in COP over the performance of the original system, which is consistent with the expected system improvement resulting from an evaporator with an 8 % larger capacity. The achieved improvement in RTU performance requires no additional material cost since it only involves changes to the refrigerant circuitry.

Keywords: Air Velocity Profile, Roof Top Air Conditioning Unit, Particle Image Velocimetry (PIV), Circuitry optimization

Acknowledgement

This study was sponsored by the United States Department of Energy, Building Technologies Program under B&R Code BT0302000-05450-1004215-Sp Cond & Ref R&D with project managers Antonio Bouza and Bahman Habibzadeh. In kind support was provided by Jake Rede, Vince Barone, and Dan Arnold of the Nordyne Corporation. Dr. W. Vance Payne and Mr. John Wamsley provided assistance with test setup and operating the environmental chambers.

Table of Contents

List of Figures

List of Tables

INTRODUCTION

Finned-tube heat exchangers are the predominant type of heat exchangers used in comfort cooling applications. Finned-tube heat exchangers are generally made up of a bundle of several dozen connected tubes. As refrigerant flows through each of the tubes, heat is transferred between it and the air flowing along the outside of the tube. The performance of the heat exchanger as a whole is the aggregate performance of every tube in the bundle. The heat transfer performance of each individual tube is influenced by many parameters including the tube and fin geometries; the refrigerant temperature, mass flux, and local quality; and the local air velocity, temperature, and humidity. The local air velocity is one of the most important parameters because it dictates the amount of air that is available for heat exchange and it influences the local air-side heat transfer coefficient. To this end, the distribution of the air incident on the heat exchanger has a profound impact on its overall performance, since it characterizes the velocity of air at each tube location in the bundle.

There has been long standing interest in air flow distributions through finned tube heat exchangers. The first documented study was that of Fagan (1980), which examined the effects on small heat exchangers used in room air conditioners. His study showed that typical air maldistributions commonly result in quite large velocity variations and that the impact on performance is significant. Chwalowski et al. (1989) later showed similar results indicating large air velocity maldistributions and went on to demonstrate as much as a 30 % variation in capacity for a given evaporator when subject to different air flow distributions. Payne et al. (2003) demonstrated that air-side non-uniformity can impose a significant reduction in heat exchanger capacity, as much as 30% in the extreme cases, which agrees with the earlier work by Chwalowski.

The current state-of-the-art for measuring air-side velocity distribution is by traversing a hot wire anemometer or pitot tube. Although this method is simple and low cost, it is cumbersome and has high measurement uncertainty both due to precision and the user's ability to maintain the position and orientation of the sensor in the exact location of interest. Furthermore, the probes for these tools are obstructive to the flow field. There are a few optically based methods available, which eliminate disruptions of the flow by the sensor probe. Particle Image Velocimetry (PIV) is used in this study since it provides the ability to characterize large sections of the flow field.

The traditional tendency is to design finned tube heat exchangers with the assumption of a uniform air velocity profile or implement a refrigerant circuitry that would offer some degree of robustness with non-uniform air distribution (Kaern, 2013). Because new tools are available to measure velocity distributions and design components that account for these distributions, the problem of air-side velocity distribution is beginning to get more attention. An approach described by Yashar et al. (2012), involving recent developments in machine learning as incorporated into the Intelligent System for Heat Exchanger Design (ISHED), has demonstrated that the capacity degradation due to air-side non-uniformity can be mitigated by designing the optimal refrigerant circuitry for the actual air flow distribution seen in the system. Furthermore, optimizing the refrigerant circuitry designs using machine learning based on knowledge of the air distribution can result in significantly improved capacity while simultaneously reducing the size, cost, and amount refrigerant charge (Domanski et al., 2010).

The objective of this study is to experimentally demonstrate the system performance improvement achieved by optimizing the evaporator's refrigerant circuitry. In the first part of this study, the in-situ air velocity distribution was measured using PIV. This detailed mapping was then used as input to the optimization tool ISHED to determine a better performing refrigerant circuitry. Next, a prototype of an evaporator with the optimized circuitry was built, installed in the system, and the predicted performance improvement was verified by laboratory measurements.

EXPERIMENTAL SETUP

This section discusses the roof top test unit, the modifications that were made to the unit in order to perform the air flow distribution measurements, and the measurement instrumentation and methods used to obtain all of the data.

2.1 Roof Top Air-Conditioning Unit

The system examined in this study is a small commercial roof top unit (RTU). The unit is an air-to-air R410A air conditioning system and has a nominal capacity of 7.5 Tons (26.4 kW). It consists of two separate vapor compression systems so that it can operate in stages under part load conditions or with both stages operating to meet full load conditions. Each stage has its own compressor, condenser and expansion valve; the unit has a single evaporator slab that is shared by the two stages with each stage utilizing half of the available heat transfer area. The two stages are designated by the names Stage 1 and Stage 2 throughout this document.

A cutaway schematic of the unit is shown in Figure 2.1.1. The unit is sectioned into three compartments. The condenser compartment (left side) consists of two condensers and a single fan that draws air from the outside through the condensers. The right side of the unit is the air discharge compartment, which delivers conditioned air to the building. The middle evaporator compartment is the focus of this study; it is where the building air is conditioned and blown into the discharge compartment.

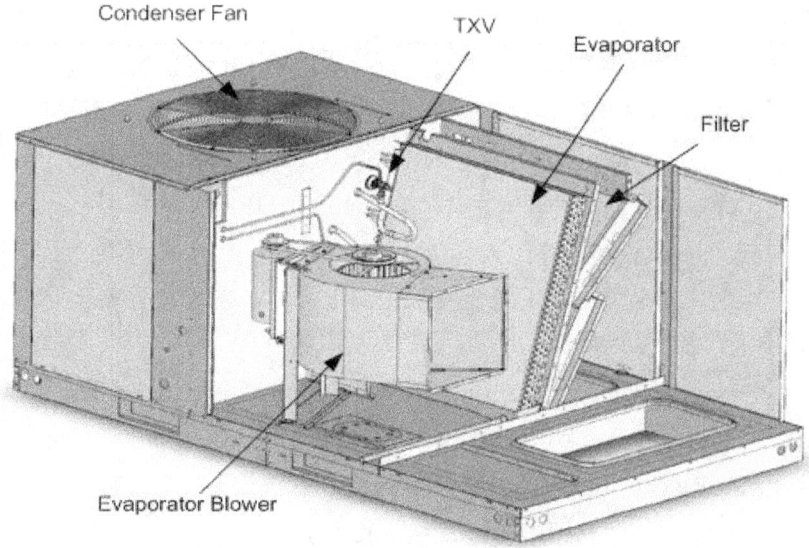

Figure 2.1.1 Schematic of Test Subject

Figure 2.1.2 shows a side view of the middle compartment. The building air enters this compartment through an intake port (not shown) located at the bottom right in Figure 2.1.2. Included in this compartment are a set of air filters, an evaporator fed by two separate refrigerant loops, a condensation collecting tray, and a blower that circulates air through the compartment and out to the discharge compartment.

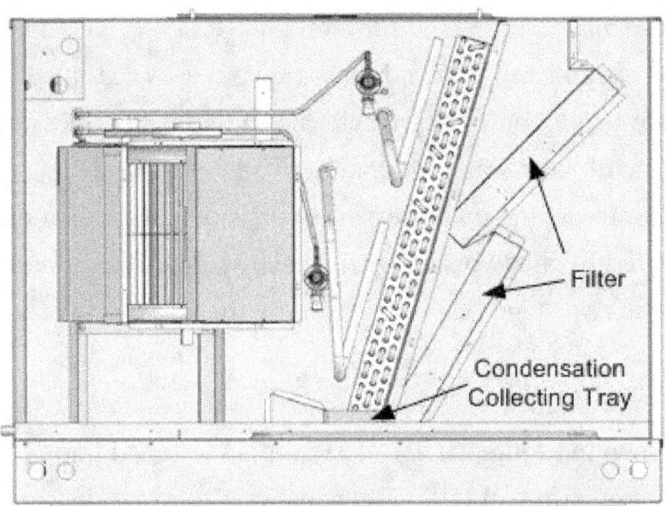

Figure 2.1.2 Middle Compartment of Test Subject

The evaporator is slanted at an angle of 15 degrees from the vertical and is made up of 144 tubes located in four depth rows with louvered fins. The tubes have an outside diameter of 9.52 mm (3/8") and are spaced 25.4 mm (1") apart along the height of the heat exchanger. The depth rows were layered in a staggered configuration and are spaced 22 mm (7/8") apart. The overall dimensions of the evaporator are 914 mm (36") high, 864 mm (34") wide, and 101.6 mm (4") thick. The total heat transfer area provided by the evaporator is divided amongst the two stages of the system. In total the evaporator is divided into 16 circuits with 8 circuits used by each stage; each circuit consists of either 8 or 10 tubes.

2.2 System Modifications

Two aspects of the test unit were modified for the purpose of experimentation. The first modification was to the condensing section of the unit. The RTU was designed to operate as an air-to-air unit, but was modified to be an air-to-water unit to facilitate testing. Each finned-tube condenser was replaced with a brazed-plate heat exchanger, which was connected to a cold water loop with variable water temperature and flow rate in order to control the refrigerant condensation. Replacing the air-cooled condensers with water cooled heat exchangers allowed measurement of the system performance while only operating a single environmental chamber.

The second modification to the system involved the RTU's enclosure. Several of the insulated metal panels that made up the walls of the enclosure were removed and replaced with 6.35 mm (1/4") thick transparent acrylic panels. These panels provided the visual access necessary to measure the air flow distribution inside the enclosure with optical techniques. Figure 2.2.1 shows a picture of the RTU from the top looking downward through one of the acrylic panels. The evaporator and blower are visible through this panel.

Figure 2.2.1 Visual Access to RTU's Evaporator

2.3 Test Apparatus and Instrumentation

The test apparatus is described in three parts. The first portion of the apparatus is the refrigerant loop shown in Figure 2.3.1. The unit is comprised of two separate refrigerant flow loops, one for each stage of cooling. Each stage has its own compressor, condenser, thermostatic expansion valve (TXV), and uses half of the heat transfer refrigerant tubes in the interlaced evaporator which are distributed across the total heat transfer area of the evaporator.   Figure 2.3.1 shows the position of the points along the refrigerant loops where the refrigerant properties were measured.   The stage 1 loop is described by the refrigerant exiting the stage 1 compressor at point 1A, and then condensed in a water cooled heat exchanger.   The liquid refrigerant exiting the stage 1 condenser flows through a coriolis mass flow meter and then onto point 1B, just before entering the enclosure for the RTU.   Once inside, it is flashed by passing through the stage 1 TXV.   After expansion, the refrigerant is distributed amongst 8 different circuits of the evaporator, which exit the evaporator through a header tube at point 1C, and then continue out of the enclosure and back to the stage 1 compressor.   The stage 2 loop is identical, and the measurement positions are likewise labeled 2A-2C.

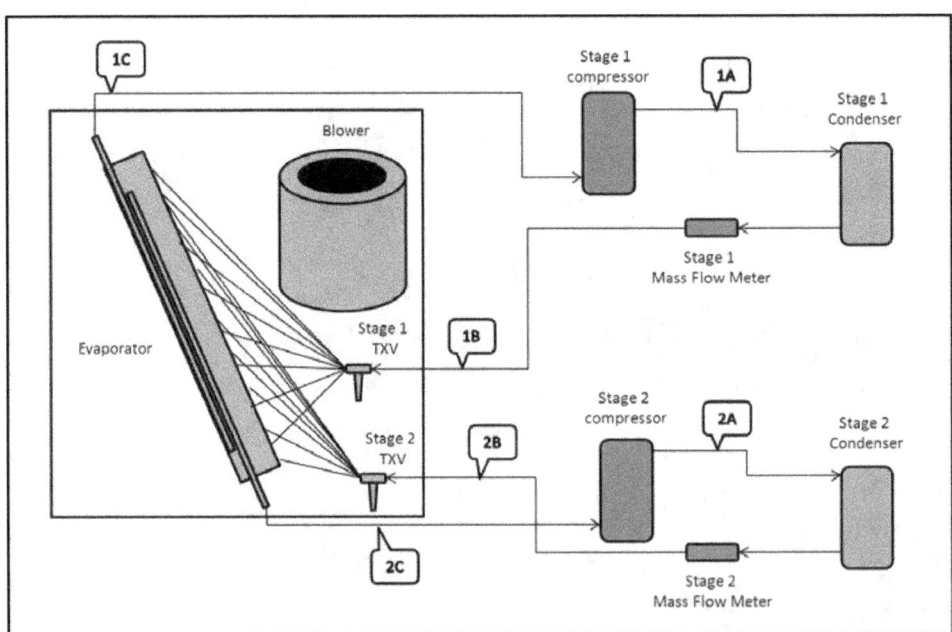

Figure 2.3.1 Refrigerant Flow Loop

The refrigerant pressure at each of these points was measured using pressure transducers with a 0 to 3447 kPa (0 to 500 psig) range and uncertainty of ±0.13 full scale at 95 % confidence.   The refrigerant temperature was measured using T-type thermocouples probes immersed in the flow, calibrated to 0.15 K (0.3 °F) at 95 % confidence.   We also measured the refrigerant temperature at several of the heat exchanger return bends using thermocouples adhered to the outside of the select tubes.   The refrigerant mass flow rate for each refrigerant loop was measured using a coriolis mass flow meter calibrated to ±1 %

at 95 % confidence.   The capacity of the RTU could be readily calculated using the refrigerant enthalpy method based on the mass flow rate and the difference between the liquid line and suction line enthalpy for each stage.   The electrical power input to the blower and each compressor was measured using three separate power analyzers calibrated to ±0.25 % at 95 % confidence.

The second portion of the apparatus is the water flow loop shown in Figure 2.3.2.   The purpose of this loop is to control the flow rate and temperature of the water that is supplied to each condenser of the refrigerant loop.   Water circulates between the condensers and a water-to-water vapor compression chiller which rejects heat to the laboratory chilled water supply.   The temperature of the water supplied to the condensers is set by the controls on the chiller.   After water exits the chiller it is divided into two paths, each one being routed to one of the condensers.   The volumetric flow rate of water is measured downstream of the split point for each path.   A needle valve is located just upstream of each condenser to control the flow rate through each path.   The water temperature is monitored at the inlet and exit of each condenser.   The chiller is capable of delivering up to 1.4 l/s (22 GPM) of water at a constant temperature by removing up to 24 kW (82 kBTU/h) of heat.

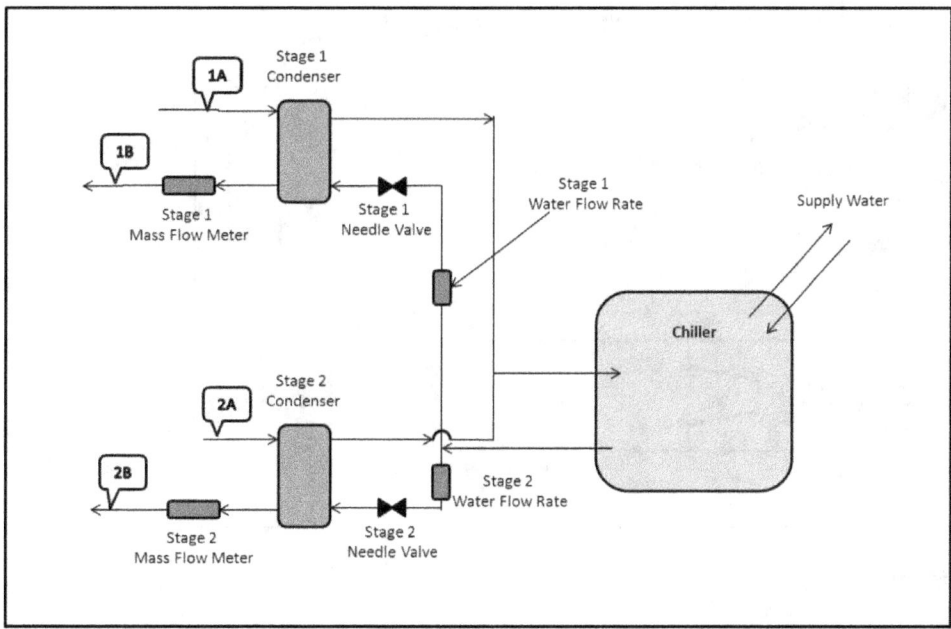

Figure 2.3.2 Water Flow Loop

The third portion of the apparatus is the air flow loop shown in Figure 2.3.3.
Conditioned air is supplied to the apparatus by controlling the conditions in the
laboratory environmental chamber. The air temperature and humidity are measured at
the inlet portion of the RTU using a grid of 25 thermocouples and a capacitive relative
humidity sensor. The conditioned air enters the RTU and passes through the evaporator
where it is cooled and dehumidified. Then it enters the blower and flows through an
insulated duct to an air flow measurement section. Upon entering this straight section of
duct, the air is mixed, straightened, and then temperature and humidity are measured.
The temperature and humidity are measured using a thermocouple grid and sampling tree
connected to a chilled mirror hygrometer. In addition, the junctions of a 9 junction
thermopile are connected to the thermocouple grids located at the air intake to the RTU
and the air flow measurement section to measure the sensible temperature difference.
The air then passes through a nozzle to measure the flow rate and then onto a booster fan
(not shown) connected at the measurement section exit before being discharged back into
the environmental chamber.

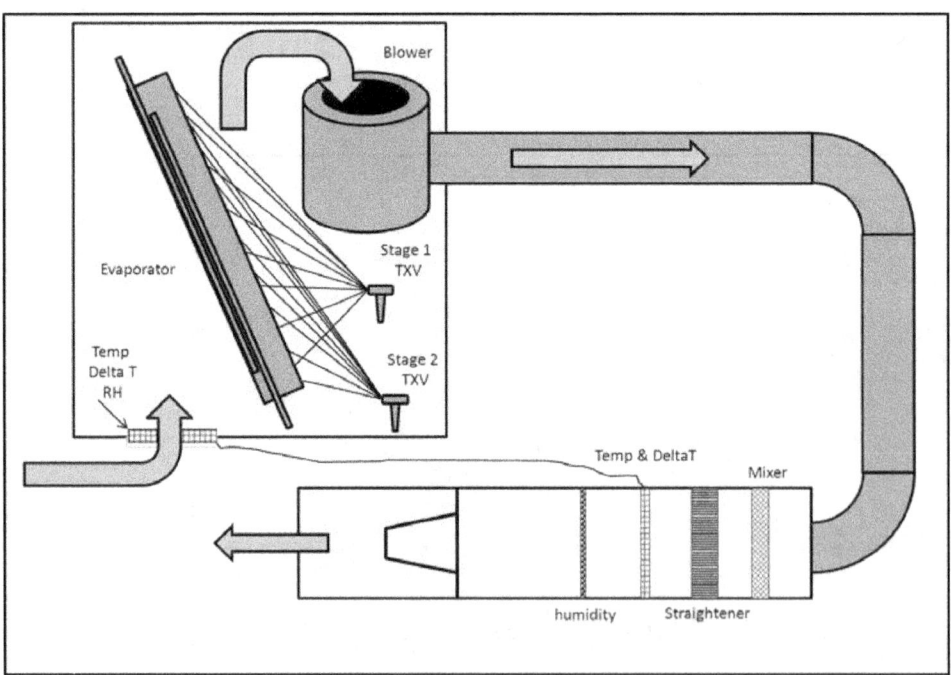

Figure 2.3.3 Air Flow Loop

The T-type thermocouples used to measure the air temperature at each point were
calibrated to 0.15 K (0.3 °F) at 95 % confidence and the thermopile had an uncertainty of
0.8 % at 95 % confidence. The air flow rate was measured using a nozzle with a

diameter of 254 mm (10") in accordance with ANSI/AMCA 210 (1985). The pressure difference across the nozzle was measured with an electronic differential pressure transducer with an uncertainty of ±0.25 % at 95 % confidence. The evaporator capacity was calculated using the air enthalpy method according to ASHRAE Standard 37-2009.

The entire apparatus, shown in Figure 2.3.4, is located in a conditioned environmental chamber which is capable of maintaining the air conditions between 10 °C (50 °F) and 60 °C (140 °F) dry bulb and relative humidity within 2 % of the target setpoint.

Figure 2.3.4 Photo of Test Apparatus

2.4 PIV Air Flow Measurement System

Particle Image Velocimetry (PIV) works on the basis of tracking the motion of particles entrained in the flow field. These "seed" particles act as markers within the flow field whose displacement can be mapped between two points in time. As the seed particles move through the test section they are illuminated by a series of laser light sheets. The laser sheets are oriented in such a way that the illuminated plane is aligned to the main flow direction within the test section; therefore particles moving downstream will remain within the illuminated plane between successive light sheet pulses. A camera is used to capture images within this plane and therefore records the distance traveled by the seed particles during the time between the light flashes, velocity is extracted from this information. Figure 2.4.1 shows a schematic of a basic 2D PIV setup.

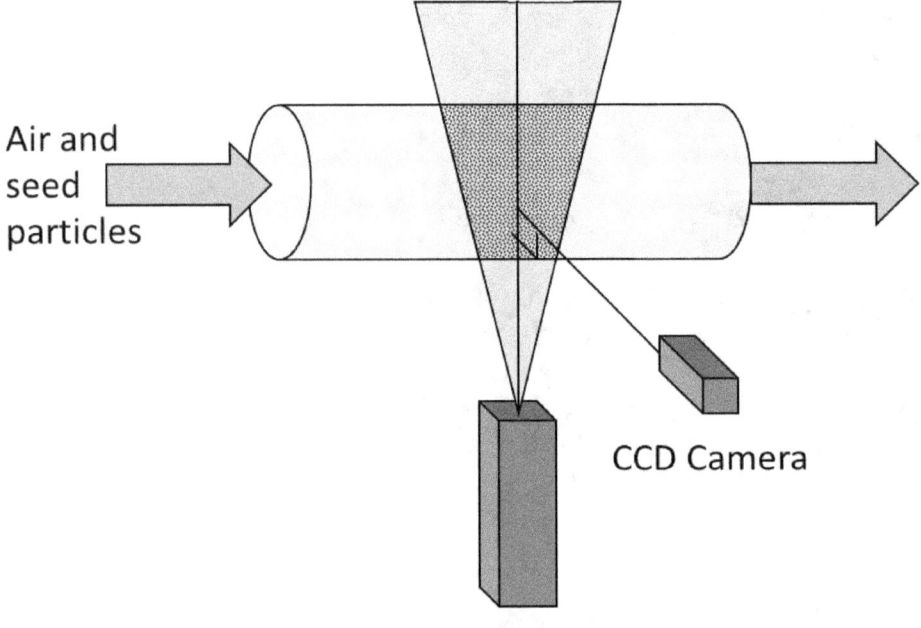

Figure 2.4.1 Schematic of a Basic 2D PIV Setup

The PIV measurement system includes of a pair of Class IV pulsed lasers outfitted with a sheet forming optical lens, a double framed Charged Coupled Device camera, a theater style fog generator, and a Programmable Timing Unit (PTU) controlled by personal computer. A detailed description of the measurement equipment can be found in Yashar et al. (2007).

Figure 2.4.2 shows the layout for the PIV equipment as set up to measure the air flow leaving the evaporator. In this configuration, the lasers are positioned on a tripod high above the top of the test unit. The laser light sheets pass through the acrylic window at the top of the test unit and illuminate planes perpendicular to the exit surface of the evaporator. The camera is set up on the side of the test subject and its view faces the light sheet through a second acrylic window. The operation of the camera and the lasers are synchronized by the PTU, and data is reduced by mapping the motion of particles entrained in the flow within the illuminated plane between successive laser pulses. Figure 2.4.3 shows the projected laser and image-capturing camera.

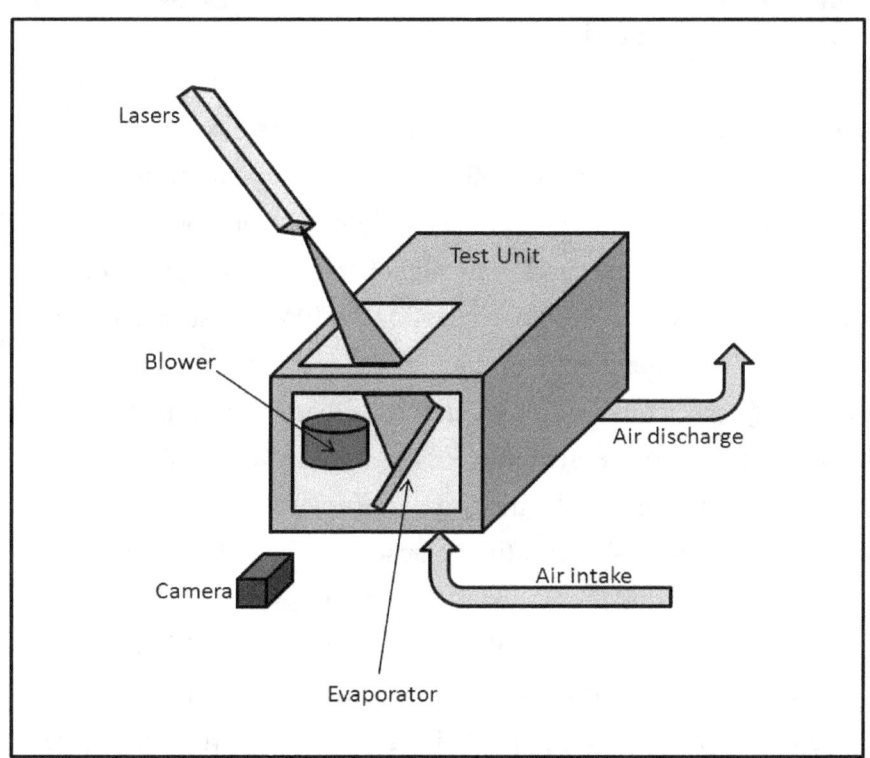

Figure 2.4.2 Layout for PIV Measurements

 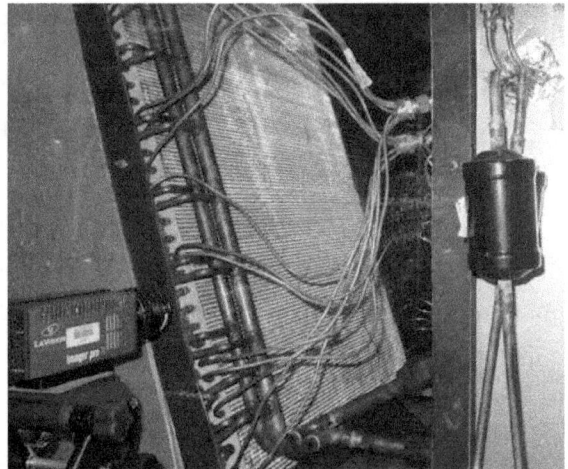

Figure 2.4.3 Projected Laser Light Sheet and Camera

The PIV system was aligned and calibrated prior to every measured data set at each specific measurement location. The detailed process for alignment and calibration was the same as that in previous study by Yashar et al. (2007). After completing the alignment and calibration process, the entire environmental chamber was filled with glycerin-based theater fog, which served as seed particles for the PIV measurements. The camera then captured a series of 100 image pairs of the seeded air flow in the measurement area. Each image pair consists of two pictures separated in time by (160 to 750) µs, depending on the data set. Successive image pairs were taken every 67 ms. The velocity field was calculated by tracking the motion of particles from each image to its paired image, which resulted in a series of 100 time elapsed vector fields. These vector fields were then averaged in time to dilute the unsteady components of the turbulent flow structures and produce a more steady-state representation of the flow field. Figure 2.4.4 shows one snapshot of PIV measurements of the air particles flowing from left to right. Figure 2.4.4 (a) shows a camera image of an illuminated plane of seed particles; Figure 2.4.4 (b) shows the velocity vector field computed from a pair of images.

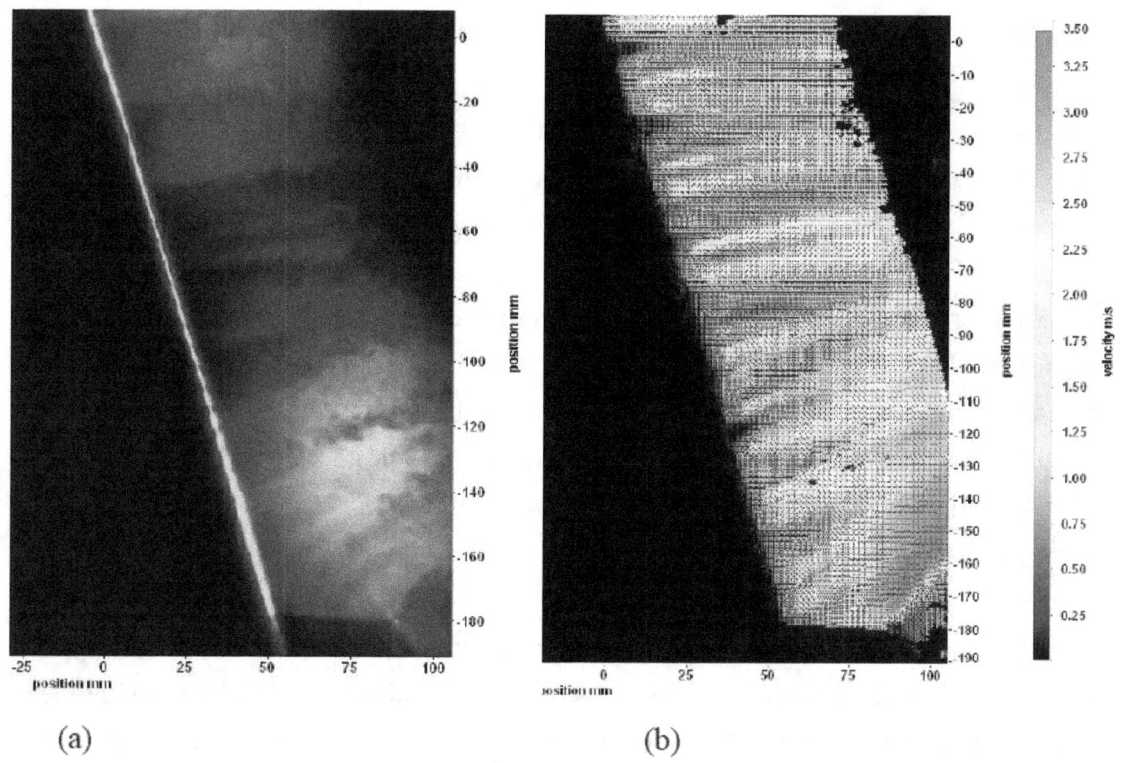

(a)   (b)

Figure 2.4.4 PIV Measurement Data: (a) Seed Particles Exiting Evaporator and (b) Vector Field Calculated from Particle Motion

3. SYSTEM TESTS WITH ORIGINAL EVAPORATOR

A series of tests were performed to characterize the performance of the unit operating with the original evaporator. To facilitate the performance measurement, the RTU's air-to-refrigerant finned-tube condensers were replaced with water-cooled brazed-plate heat exchangers. The tests of the modified system were performed in such a way that the evaporator operated exactly as it would in the original configuration. Since the unit is a production model of a commercially available air conditioner, its performance was tested by its manufacturer in accordance with AHRI Standard 340/360 (2007) as part of the certification process. The manufacturer assisted with this project by providing their test data for this unit, which served as a basis for the evaporator operation during the laboratory measurements in this study.

This rating standard requires operating two environmental chambers simultaneously; one to supply indoor conditions and one to supply outdoor conditions. The indoor chamber must supply 1.42 m^3s^{-1} (3000 CFM) of conditioned air at 26.7 °C (80.0 °F) dry bulb and 19.4 °C (67.0 °F) wet bulb to the evaporator and the outdoor chamber must provide an environment for the condenser at a temperature of 35.0 °C (95.0 °F). While the manufacturer performed these tests, they recorded enough data to calculate the refrigerant state properties at the inlet of each TXV and the outlet of the evaporator. The refrigerant charge and the condenser water temperature and flow rate were adjusted in order to match each of the parameters shown in Table 3.1 during the performance tests of the baseline system.

Table 3.1 Target Refrigerant Conditions During Baseline Performance Tests

	TXV Inlet			Evaporator Outlet		
	Temperature (°C)	Pressure (kPa)	Subcooling (°C)	Temperature (°C)	Pressure (kPa)	Superheat (°C)
Stage 1	38.1	2690	6.2	12.5	1120	1.4
Stage 2	37.4	2760	8.1	11.1	1120	3.2

Throughout this study, the expectation is that the overall improvement in system capacity would be fairly small because of the nature of the system's operation. Domanski (1988) provides a method of predicting the performance of a system with a replacement evaporator in a residential split system while keeping all other components the same.

This model generally predicts a change in system capacity that is on the order of 1/3 of the difference in capacity between the new and original evaporator measured at the same saturation temperature because the system must rebalance itself at a different set of saturation temperatures in the evaporator and condenser. For this reason, best efforts were taken to minimize all of the measurement uncertainty.

During the tests conducted as part of this study, the capacity measured by the air enthalpy method was 2 % - 3 % lower than the capacity measured by the refrigerant enthalpy. The main contribution for this difference is conduction of heat through the enclosure walls, which is enhanced due to the fact that insulation was removed from the RTU's enclosure in order to outfit it with the acrylic panels and provide visual access inside during operation. Other factors, including air infiltration into the negative pressure compartment of the enclosure, make significant contributions to this difference. For this reason it was determined that the refrigerant side capacity measurements would provide a better basis for this work than the air-side capacity measurements, which are required by the standard test method, since it is a direct measurement of the evaporator's performance.

A total of eight individual measurements of the system's capacity were recorded over a two week period; relevant portions of the recorded data are provided in Appendix A. The results of these tests measurements are shown below in Table 3.2. The average capacity from these tests is (26.42 ±0.29) kW ((90±1) kBTU/h), which is within 1.8 % of the value from the manufacturer's certification data of 26.85 kW (91.6 kBTU/h).

It is interesting to examine the values obtained for the ratio of cooling capacity to compressor power, which is a measure of the Coefficient of Performance (COP). There are four different values for COP listed in the Table. COP Stage 1 is the ratio of cooling capacity from Stage 1 to the compressor power for Stage 1; it is a measure of the efficiency of the Stage 1 system. Likewise, COP Stage 2 is a measure of the Stage 2 system efficiency. These values are useful to compare the efficiency of each stage in the system, because they exclude the electrical power consumption of the shared components. The system COP w/o blower is the ratio of the total cooling capacity from both stages to the total compressor power input to the system. The last column in the table lists the System COP, which includes the electrical power input from both compressors and the blower. Since this system was modified to operate as an air-to-water unit, the system COP does not include the influence attributed to the electric energy input to the condensing section that was originally built with the system.

Table 3.2 Refrigerant Side Capacity and COP Measurements for System with Original Evaporator

Test Number	Capacity stage 1 (kW)	Capacity stage 2 (kW)	Capacity total (kW)	Stage 1 Comp power (W)	COP Stage 1	Stage 2 Comp power (W)	COP Stage 2	System COP w/o blower	Blower Power (W)	System COP
1	13.21±0.21	13.13±0.24	26.34±0.32	2927±7	4.51±0.07	3069±5	4.30±0.08	4.39±0.05	1036±27	3.75±0.05
2	13.18±0.20	13.20±0.23	26.38±0.30	2921±3	4.51±0.07	3067±5	4.30±0.08	4.41±0.05	1014±27	3.77±0.05
3	13.23±0.20	13.24±0.24	26.47±0.31	2923±8	4.53±0.07	3069±8	4.31±0.08	4.42±0.05	1010±20	3.78±0.05
4	13.24±0.17	13.15±0.19	26.39±0.25	2935±3	4.51±0.06	3072±6	4.28±0.06	4.39±0.04	1019±28	3.76±0.04
5	13.23±0.18	13.21±0.20	26.44±0.26	2926±5	4.52±0.06	3067±3	4.31±0.07	4.41±0.04	1030±25	3.76±0.04
6	13.27±0.26	13.16±0.20	26.43±0.33	2932±6	4.53±0.09	3071±3	4.29±0.07	4.40±0.05	1023±30	3.76±0.05
7	13.31±0.18	13.17±0.21	26.48±0.28	2931±2	4.54±0.06	3054±13	4.31±0.07	4.42±0.05	1018±27	3.78±0.04
8	13.25±0.15	13.19±0.21	26.44±0.26	2925±4	4.53±0.05	3050±4	4.33±0.07	4.43±0.04	997±19	3.79±0.04
Average	**13.24±0.19**	**13.18±0.22**	**26.42±0.29**	**2927±7**	**4.52±0.06**	**3065±10**	**4.30±0.07**	**4.41±0.05**	**1018±26**	**3.77±0.04**

The expressed values are shown as the average of all measurements ± the total combined uncertainty at 95% confidence as described in Appendix C.

It is important to note that the values for uncertainty are strictly based on the variance and measurement uncertainty of the measured refrigerant mass flow rates, pressures, and temperature data and do not consider the uncertainty of the calculated values of refrigerant enthalpy inherent to the subroutines in REFPROP (Lemmon et al., 2010), which are typically on the order of 1 % (Lemmon and Jacobsen, 2004). This was left out of the uncertainty calculations because these performance measurements will be compared to performance measurements of the system with the optimized evaporator, which will include the same uncertainty components and these components will not influence the comparison of the two systems. By accepting this approach, the measurement uncertainty for comparative purposes is close to 1 %, which is significantly better than that which could be obtained had the air-side capacity measurements been used. Example calculations for the measurement uncertainty are provided in Appendix C.

4. AIR FLOW DISTRIBUTION TESTS

The air flow distribution at the evaporator was measured while operating the unit under performance test conditions. In order to ensure that the heat exchanger was operating in a steady manner with an adequate level of condensate on the outside surface, the unit was operated for a minimum of 30 min between calibrating the PIV measurement system and collecting the air flow distribution data. Air velocity data was collected at six different lateral positions along the surface of the heat exchanger and the air flow distribution exiting the evaporator along each of these planes was determined from the measurements. The positions that were queried were 70 mm, 210 mm, 330 mm, 450 mm, 560 mm, and 740 mm from the edge of the heat exchanger, as indicated in Figure 4.1.

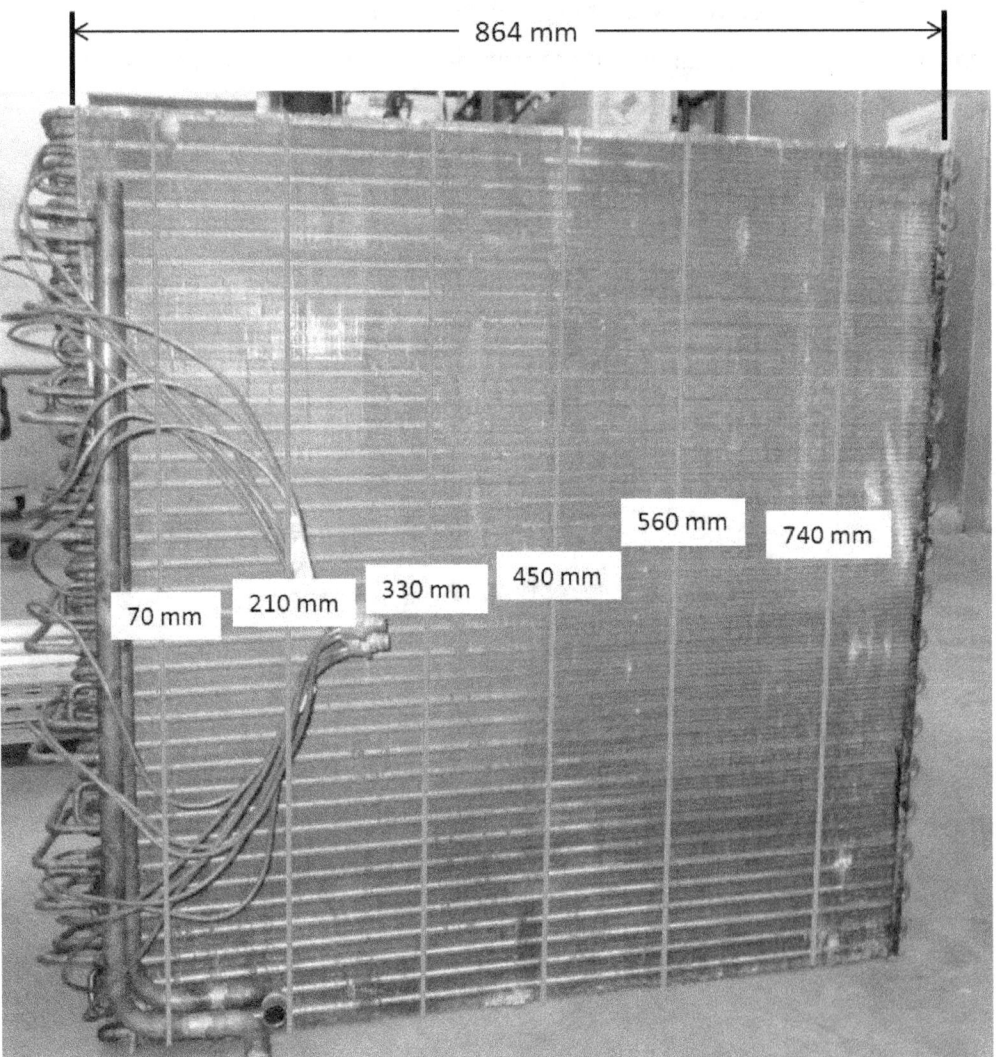

Figure 4.1 Evaporator with Marked Air Velocity Measurement Planes

Figure 4.2 shows a representative set of PIV data collected during these tests. These data were taken at the position of 330 mm from the edge. The data was collected in seven separate scans and is assembled piecewise with the first picture, labeled V(a), at the top of the heat exchanger and V(g) at the bottom, and aligned to form a complete velocity profile data set.

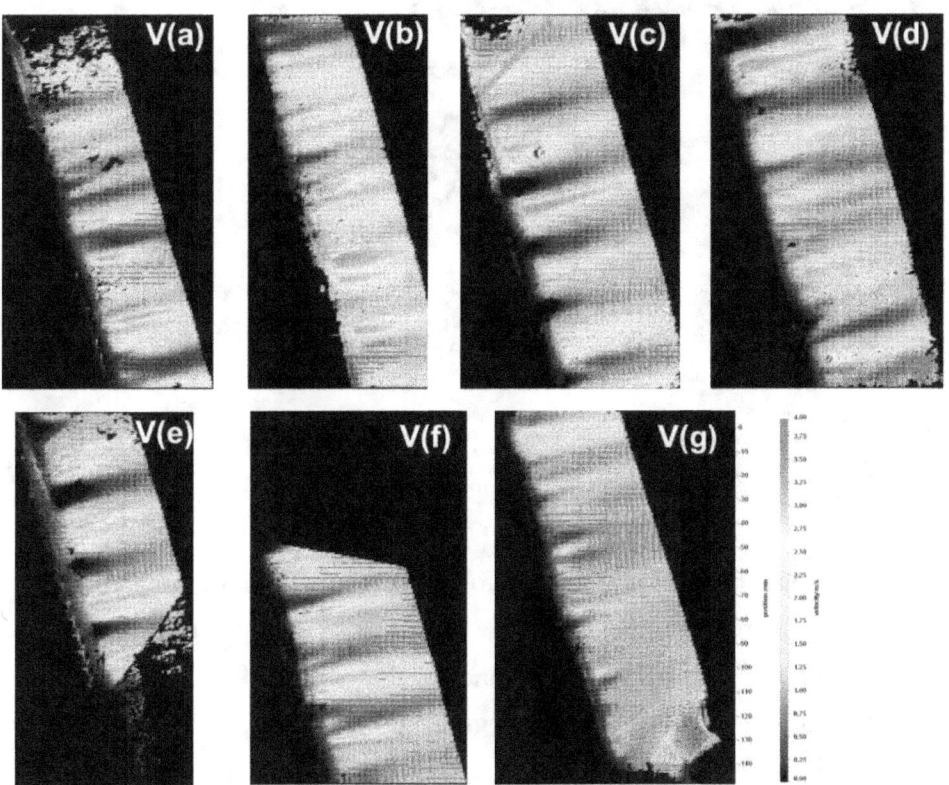

Figure 4.2 Velocity Vector Field at 330 mm

Notice that these figures include patterns with several bands of high and low velocities exiting the evaporator. There are two aspects that contribute to these patterns. First, the pattern stamped into the heat exchanger's fins cause a distinct pattern of small jets. Secondly, the tubes in the last depth row of the heat exchanger cause larger obstructions which locally impede the flow. The patterns induced by these two factors are highlighted in Figure 4.3.

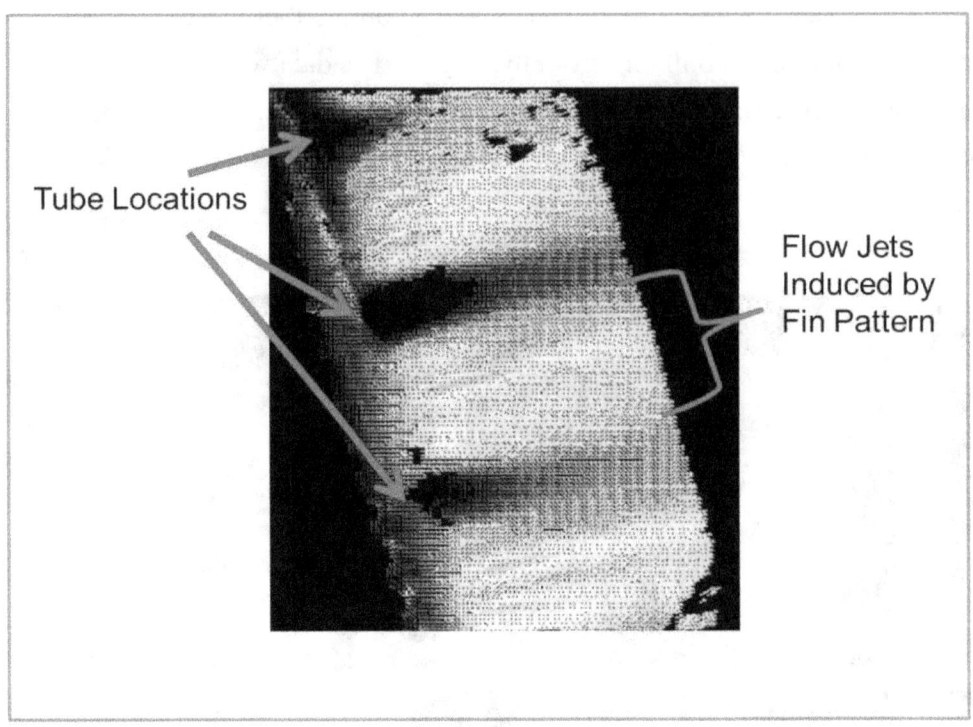

Figure 4.3 Blowup of Velocity Field at Heat Exchanger Outlet

The tube locations and jet patterns are quite useful for ensuring proper alignment of the data sets when assembling the piecewise segments.   It is, however, not information that can be used for the simulation phase of this study, and for that reason the flow jet patterns were removed from the data set by numerically integrating the measured velocity between each pair of adjacent tube locations and assigning a value of the average velocity to each midpoint between them.   Figure 4.4 illustrates the integration process.   The dashed black lines represent the measured velocity as a function of position extracted from the PIV data, the light blue vertical lines represent the locations of tubes in the last depth row of the heat exchanger, and the red markers indicate the average velocity between adjacent tube locations as determined by numerical integration of the data.

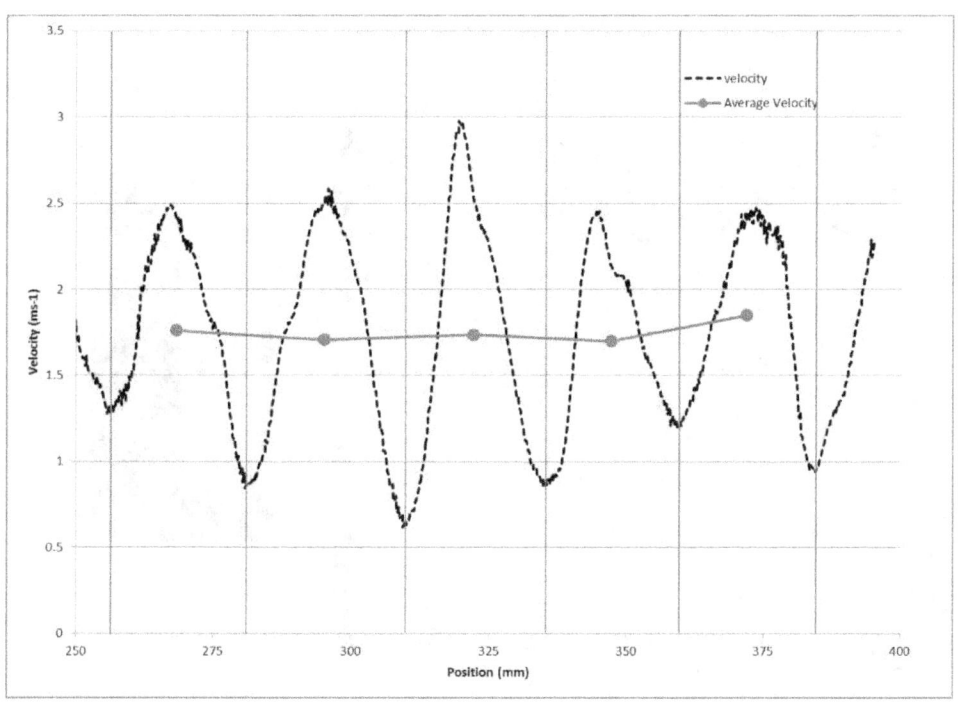

Figure 4.4 Numerical Integration of Data to Remove Periodic Patterns Associated with Tube Sites

Once velocity distribution data set was compiled for each of the 6 lateral positions, it was assembled into the flow map shown in Figure 4.5. This figure was constructed by linearly interpolating between the data collected at each of the adjacent data sets; i.e. data collected along the 210 mm and 330 mm lateral positions were used to fill in the area between them. Data for lateral positions between the heat exchanger's edge and the location of 70 mm was assumed constant for the purposes of generating the flow map; likewise for the data at lateral positions greater than 740 mm. The coordinates listed on the flow map reference the upper left-most position on the exit side of the heat exchanger shown in Figure 2.1.1 as the origin. This flow map does show several patterns that are a result of the features within the RTU enclosure; for example, the air flow near the bottom of the heat exchanger (vertical distance > 800 mm) is significantly impeded by the condensate collection pan located there and it is also impeded near the lateral midline of the map by a mounting bracket for the unit's air filter located upstream of the heat exchanger.

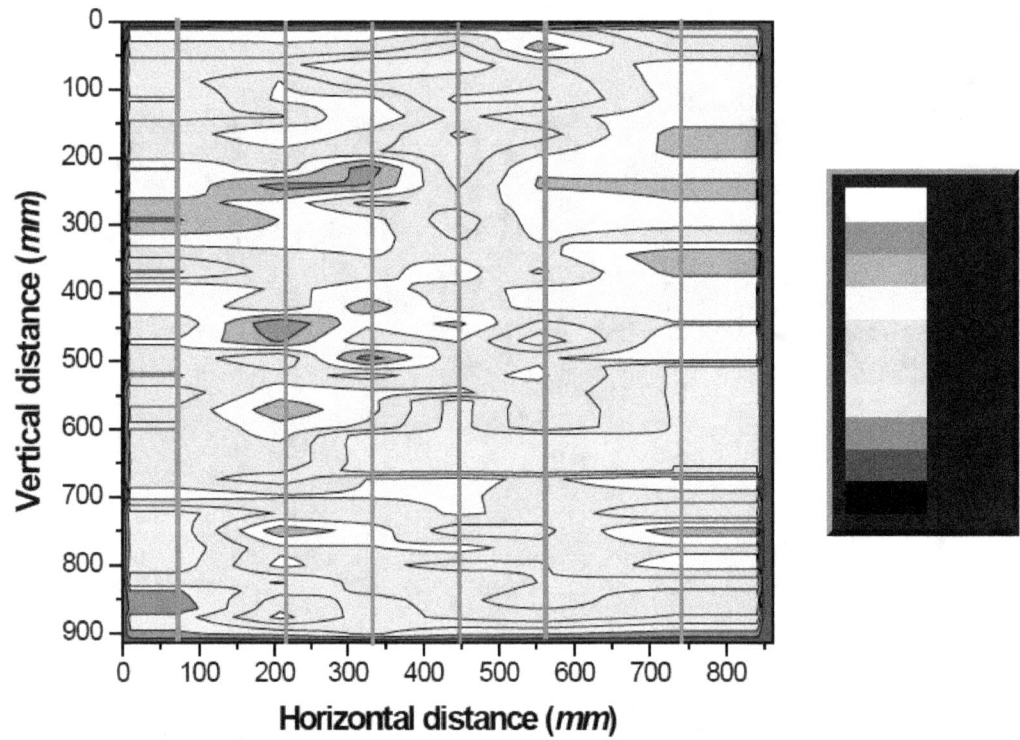

Figure 4.5 Air Velocity Flow Map

Once the flow map was complete, data was extracted from it for the simulation phase of this study.   The form of the data required as input to the simulation model, EVAP-COND, is that of a 1D distribution used as the basis for a tube-by-tube analysis. Therefore, the velocity map was numerically integrated to determine the appropriate flow distribution.   The tubes in the heat exchanger are laid out in such a way that each tube is located at a constant vertical position and span the horizontal distance shown on the map; therefore we integrated the flow map in the horizontal direction at every vertical position to generate the 1D distribution shown in Figure 4.6.   The return bends for the heat exchanger are also shown on this figure for illustrative purposes; each air velocity data point represents the average air velocity that a tube in that location would realize.

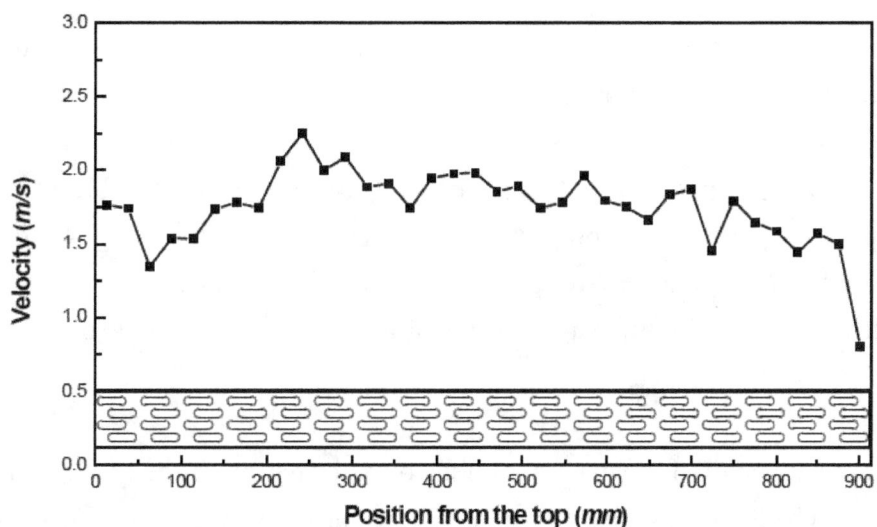

Figure 4.6 1D Flow Distribution

5. PERFORMANCE OPTIMIZATION

The next phase of the study involves the computational effort of modeling, simulating, and optimizing the performance of the evaporator based on the measurement results.

5.1 Modeling and Simulating Performance of the Original Evaporator

The first step of this phase was to prepare a model of the evaporator using NIST's heat exchanger simulation software EVAP-COND (Domanski and Yashar, 2010), which uses a first principles based, tube-by-tube approach to simulate the performance of finned-tube air-to-refrigerant heat exchangers.

EVAP-COND requires all of the general dimension information of the heat exchanger at the onset.   One challenge associated with using the current version of EVAP-COND is that the program's structure is currently limited to heat exchangers whose total number of tubes is 130 or less.   Since the evaporator used in this study has a total of 144 tubes (4 rows of 36 tubes) the model had to be modified so that the design would fit within this constraint.   The original design of the evaporator is laid out as shown below in Figure 5.1.1.   In this figure, the inlet tubes are shown as thick-walled red circles and the exit tubes are shown as thick-walled blue circles.   The return bends are shown as lines connecting the tubes, the solid lines represent return bends on the near side of the heat exchanger, and the dashed lines represent return bends on the far side of the heat exchanger.

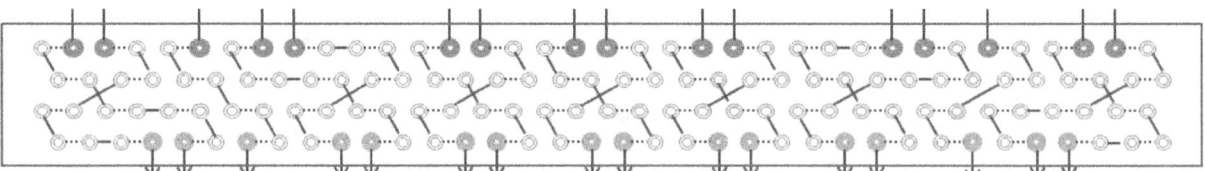

Figure 5.1.1 Sketch of Original Evaporator Design

One simplifying aspect of this design is that the return bends on the far side of the heat exchanger are patterned in such a way that they all connect adjacent tubes within the same depth row.   This pattern is called a "hairpin" pattern because each of the tube pairs resembles a large hairpin.   This pattern simplifies the manufacturing process because each hairpin is one long tube that is bent 180° at its midpoint.   Given the fact that this design connects each tube to its neighbor in the same depth row and that the air flow distribution is continuous, the layout of the evaporator simulation model was reduced by redefining each tube pair as a single tube with twice the length of the original tubes. This resulted in a reduced matrix model with the total number of tubes cut in half. The 72 tube reduced matrix representation of the evaporator is shown in Figure 5.1.2.

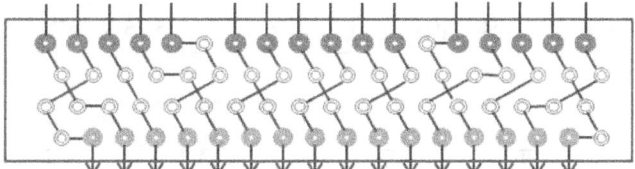

Figure 5.1.2 Sketch of Reduced Matrix Representation of Original Evaporator Design

The information required by EVAP-COND includes the number of tubes and how they are distributed in the slab; the length, inner, and outer diameter of the tubes; the tube spacing, material, and inner surface patterning; as well as information about the fins. The coil design data used for this model is shown in Figure 5.1.3.

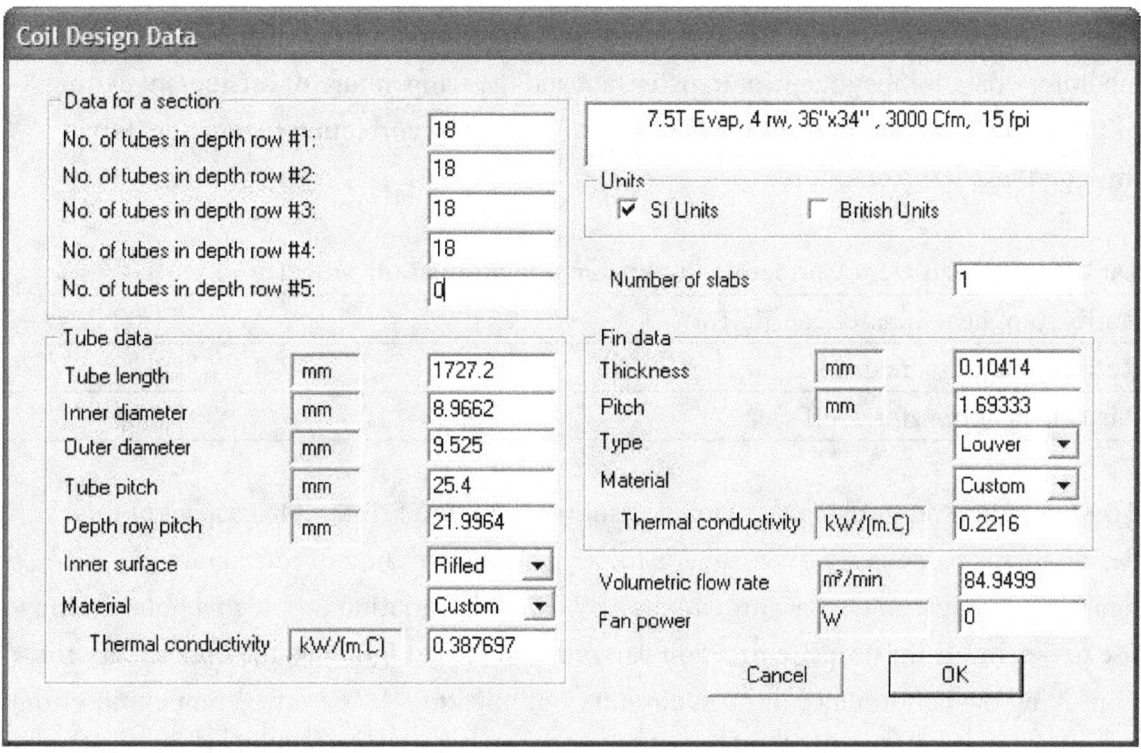

Figure 5.1.3 Coil Design Data for reduced Evaporator Design

Once the layout of the evaporator model was constructed, the refrigerant circuitry and the 1D air flow distribution were input to the model. The operating conditions measured during the laboratory tests were used for the simulations.

Table 5.1.1 Operating Conditions for Simulation Model

Refrigerant saturation temperature at evaporator outlet	11.7 °C
Refrigerant superheat at evaporator outlet	4 °C
Refrigerant mass flow rate (estimate for beginning iterative calculations)	595 kgh^{-1}
Inlet refrigerant quality	22.4 %
Air inlet temperature	26.7 °C
Air inlet pressure	101.325 kPa
Air inlet relative humidity	51.1 %

The correlations in EVAP-COND that are used to calculate heat transfer coefficients and pressure drop are typically based on measured data with significant uncertainty and are also affected by the specific patterning on the tubes and fins.   For this reason, EVAP-COND allows the user to input correction factor multipliers for the refrigerant heat transfer coefficient, refrigerant pressure drop, and air-side heat transfer coefficient. These correction parameters were adjusted until the computational result matched the laboratory data for the total heat transfer rate and the temperature of refrigerant exiting 8 of the 16 individual circuits of the heat exchanger.   The correction parameters found through these iterations were:

Table 5.1.2 Coefficient Correction Parameters for Simulation Model

Refrigerant heat transfer coefficient:	1.02
Refrigerant pressure drop:	1.20
Air-side heat transfer coefficient:	0.83

The simulation result using this model is shown in Figure 5.1.4.   One aspect of this design that is particularly interesting is the significant variation of refrigerant exit temperature between the various circuits.   While this variation provided a good metric to use to determine the proper correction parameters, it also highlights the opportunity for improving the performance through circuitry optimization.   The refrigerant exiting from the tubes numbered 9, 10, 11, 12, and 15 in Figure 5.1.4 contain high quality two-phase refrigerant, and the refrigerant exiting from each of the other exit tubes is significantly superheated in order to compensate for the fact that some of the streams contain liquid droplets.   The refrigerant quality and temperature exiting each of the heat exchanger tubes is shown in Figure 5.1.5 and Figure 5.1.6, respectively.

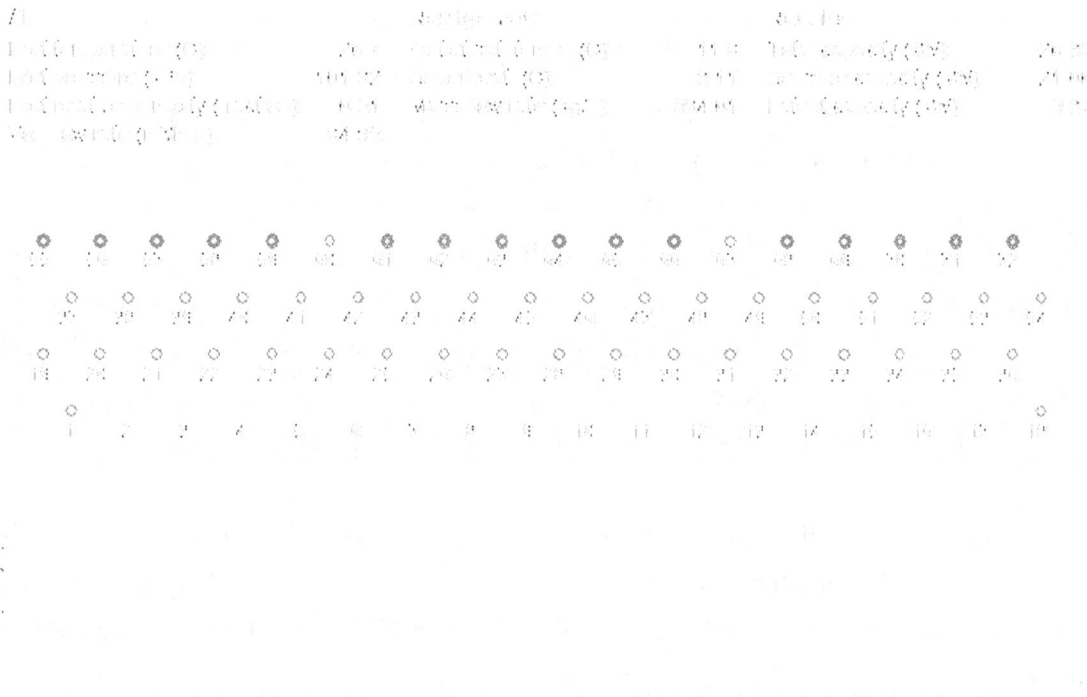

Figure 5.1.4 Simulation Result from Original Evaporator Design

Figure 5.1.5 Refrigerant Outlet Quality from Simulation

Figure 5.1.6 Refrigerant Outlet Temperature (°C) from Simulation

The original heat exchanger design consists of 16 circuits and each circuit is comprised of either 8 or 10 tubes because an even number of tubes per circuit is required to preserve the hairpin pattern. The 10-tube circuits would impart more pressure drop on the refrigerant flowing through them than the 8-tube circuits, if all other factors were equal. However, since the pressure drop through each circuit must be equal, the 8-tube circuits will generally receive a higher mass flow rate of refrigerant than the 10-tube circuits. It is important to know that the variation of the air flow also plays a significant role in the refrigerant distribution. This is because it influences the heat transfer rate on the local level, and as the refrigerant absorbs heat it changes phase from liquid to vapor. This phase change results in a significant acceleration due to the change in density, and the resulting higher velocity increases the frictional pressure drop per length of tube.

It is not surprising that all of the 10-tube circuits resulted in superheated refrigerant at their exit because the length of these circuits would cause refrigerant to be preferentially routed through the 8-tube circuits and because they have 25 % more heat transfer area than the 8-tube circuits. Three of the 8-tube circuits also resulted in a refrigerant exit condition of superheated vapor while the other five did not. The 8-tube circuits that resulted in a superheated exit condition were concentrated in locations that received more air flow than the other 8-tube circuits.

5.2 Refrigerant Circuitry Optimization

Once a model of the heat exchanger was developed, it was used as the basis for optimizing the refrigerant circuitry using the Intelligent System for Heat Exchanger (ISHED) module. Since ISHED is based on evolutionary algorithms, rather than employing calculus based techniques, independent optimization runs result in different designs. For this reason, seven independent optimization runs were performed and the best design from each one was selected for further analysis. The following parameters were used as input to control each optimization run:

Table 5.2.1 ISHED Control Parameters for Circuitry Optimization Run

Number of designs per generation	40
Number of generations	250
Minimum number of inlet tubes	12
Maximum number of inlet tubes	18

In addition, the search space was limited to include designs that had evaporator exit tubes located in the first depth row of the heat exchanger, since this is a desirable design feature to connect the exit tubes to a header.

The best out of the seven designs, with a capacity that is 7.9 % higher than the original design, is shown in Figure 5.2.1. This design contains many long, complicated crossover return bends and therefore would be extremely difficult or impossible to manufacture; however, it can be altered into a buildable design.

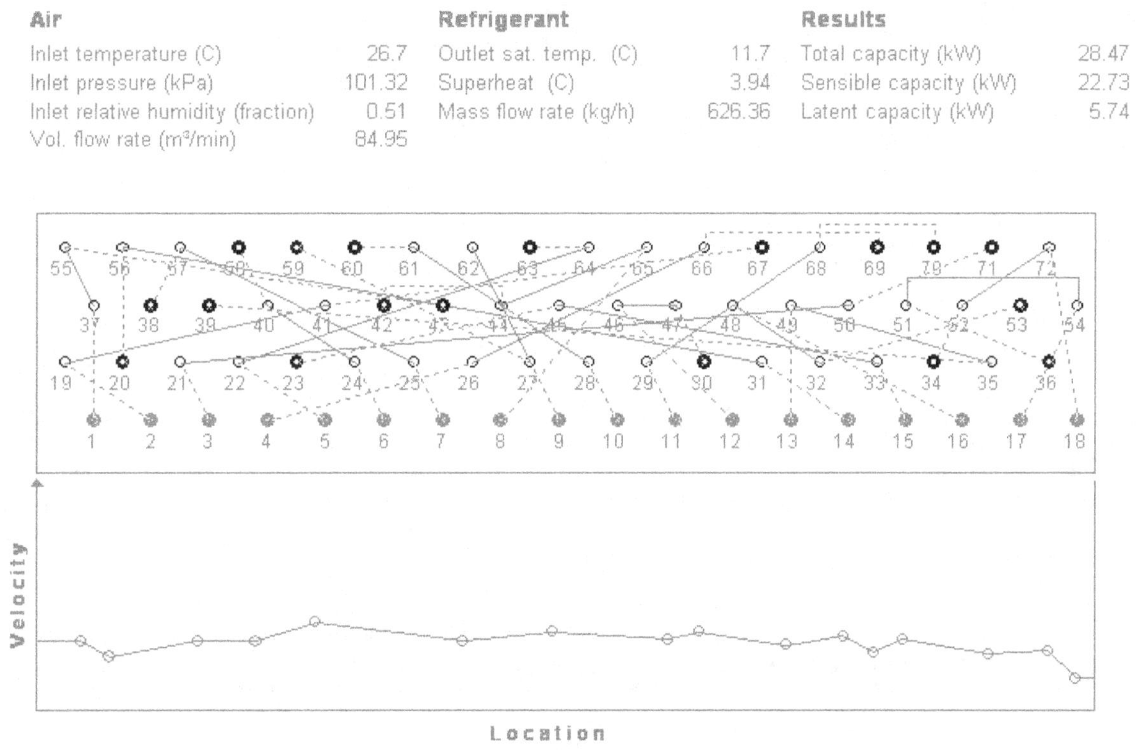

Air		Refrigerant		Results	
Inlet temperature (C)	26.7	Outlet sat. temp. (C)	11.7	Total capacity (kW)	28.47
Inlet pressure (kPa)	101.32	Superheat (C)	3.94	Sensible capacity (kW)	22.73
Inlet relative humidity (fraction)	0.51	Mass flow rate (kg/h)	626.36	Latent capacity (kW)	5.74
Vol. flow rate (m³/min)	84.95				

Figure 5.2.1 ISHED Optimized Design Before Post-Processing

The main reason that this design shows a significant improvement in capacity over the baseline design is that it does a better job of balancing the heat transfer, refrigerant pressure drop, and mass flow rate between the individual circuits. Furthermore, this design consists of 18 circuits, each with 8 tubes and each circuit providing a similar amount of heat transfer. The refrigerant pressure drop through this heat exchanger will be smaller than that through the original design because the refrigerant has more paths through it and the paths are shorter; this design will therefore accommodate a larger total mass flow rate.

This design was manually altered by swapping tube connections to untangle the cross-over bends. The resulting "cleaned" design is shown below in Figure 5.2.2, which preserves the balance between the circuits and maintains the capacity. Figure 5.2.3 shows the same design without the reduced matrix representation.

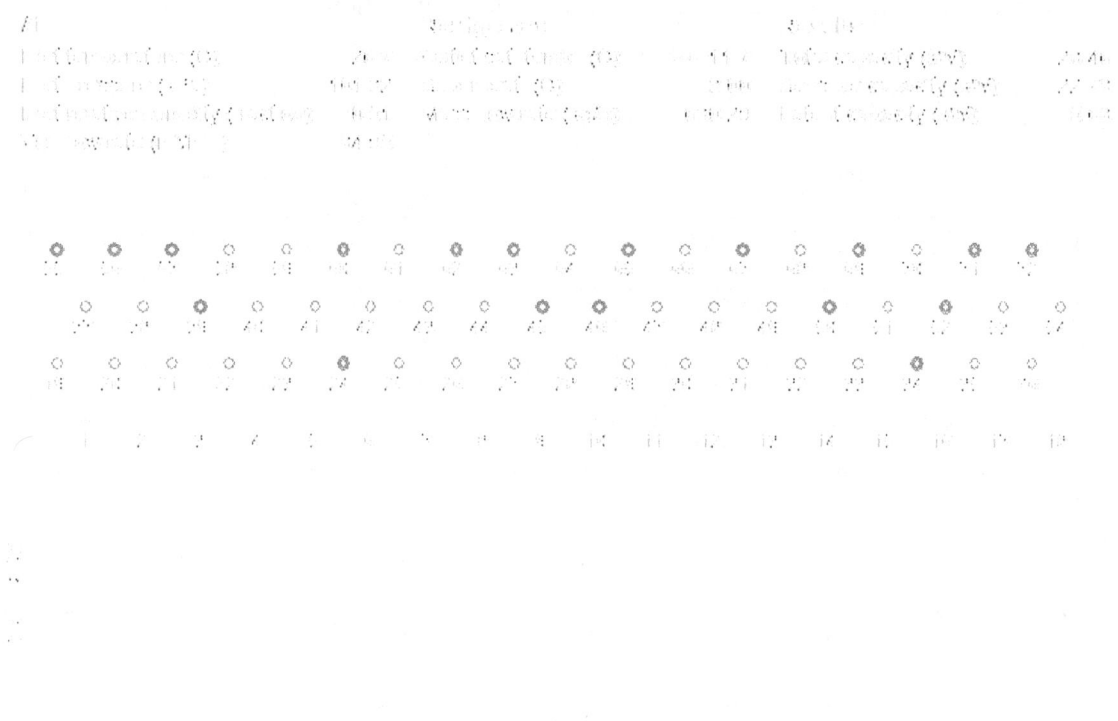

Figure 5.2.2 ISHED Optimized Design After Post-Processing

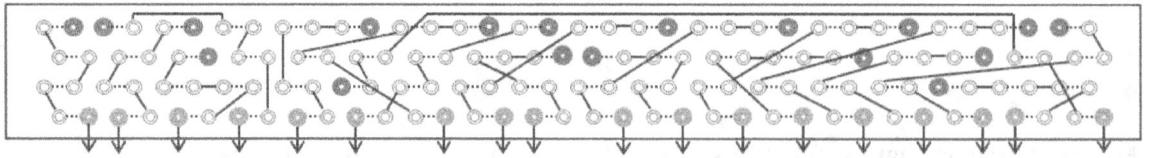

Figure 5.2.3 ISHED Optimized Design After Post-Processing, Full Representation

The next step was to group the circuits in this design into two stages. This was done with two objectives. The first objective was that of keeping an even balance between the capacity and mass flow rate between the two stages. This was accomplished by tabulating the refrigerant mass flow rate and capacity for each circuit and determining all of the possible combinations that would result in an even split between the stages. The

30

second objective was to arrange the circuits of each stage so that they were spread across the entire height of the evaporator, which will improve the performance of the system in single stage operation.   Through this step, the circuits with exit tubes 3, 4, 7, 8, 10, 13, 14, 16, and 18 were grouped into stage 1 and the remaining circuits into stage 2.

The performance of the heat exchanger in single stage operation was also simulated in order to ensure that the circuits are balanced during part load operation.   The simulation results for single stage operation of each stage are shown in Figures 5.2.4 and 5.2.5.

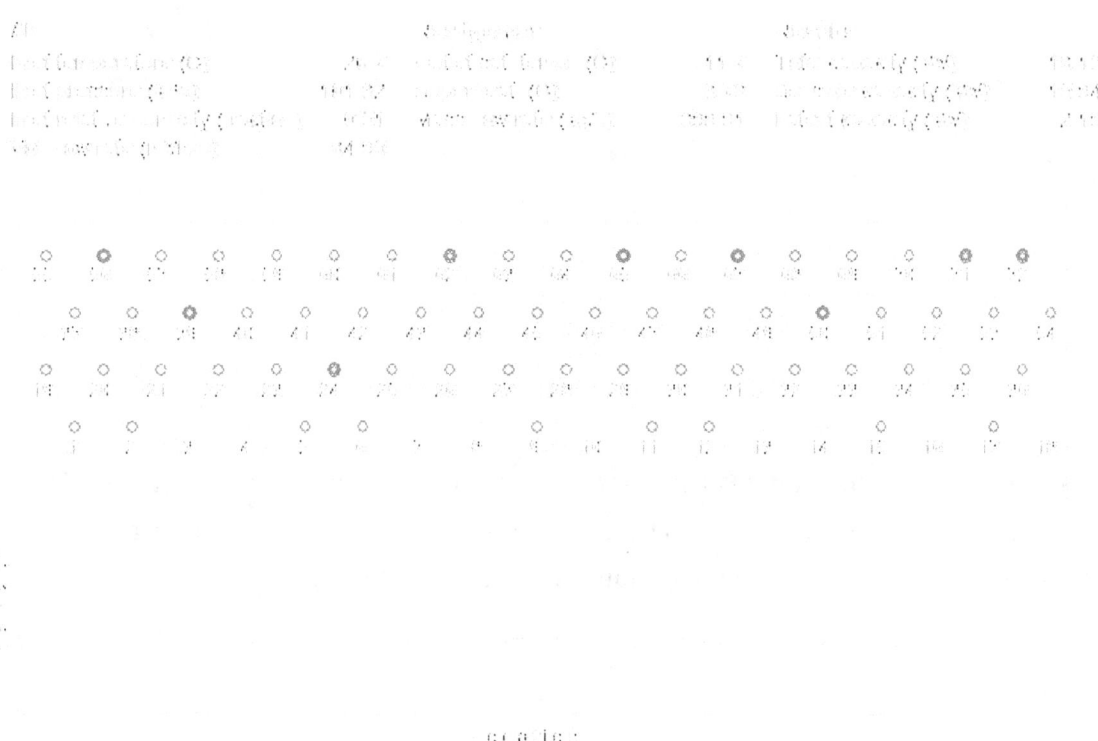

Figure 5.2.4 Simulation of Stage 1 Operation of Optimized Evaporator

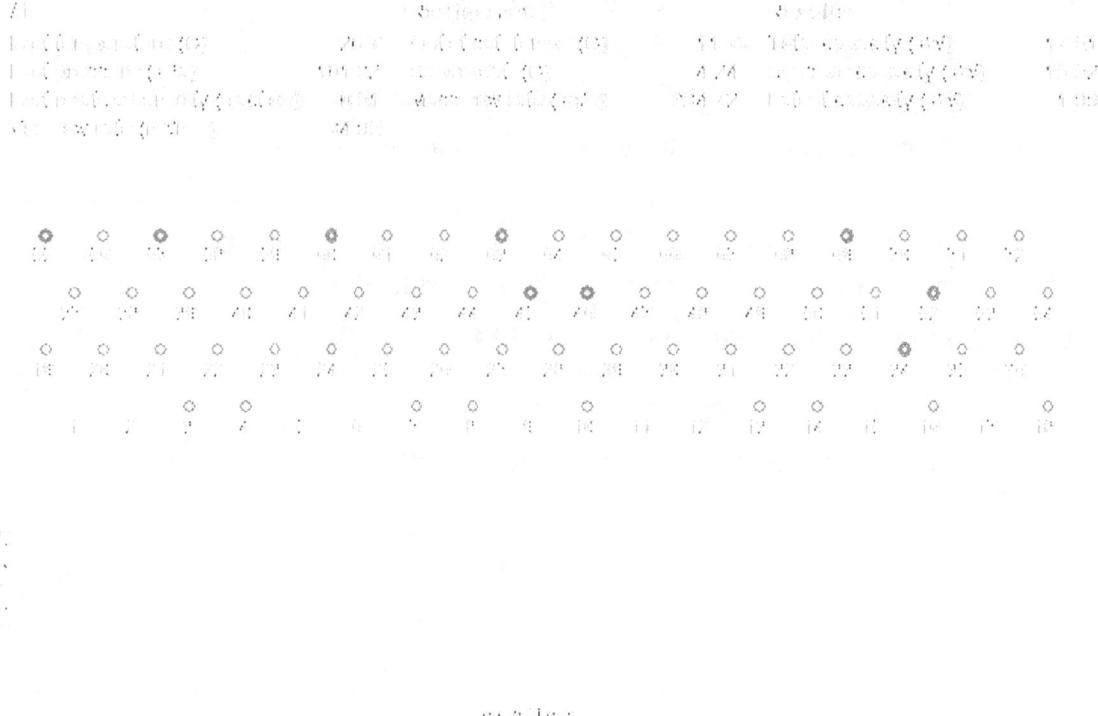

Figure 5.2.5 Simulation of Stage 2 Operation of Optimized Evaporator

The simulations for the original design show capacities of 17.42 kW and 17.11 kW for Stage 1 and Stage 2, respectively. Therefore, these results show that there is a significant improvement for each individual stage of operation.

6. SYSTEM TESTS WITH OPTIMIZED EVAPORATOR

6.1 System Setup and Test Results

After completing the simulation and optimization phase of this study, the equipment manufacturer built a prototype of the evaporator with the optimized circuitry. Figure 6.1 shows both the original and optimized evaporators side by side. While the optimized evaporator has a much more complicated circuitry design than the original evaporator, it is manufacturable.

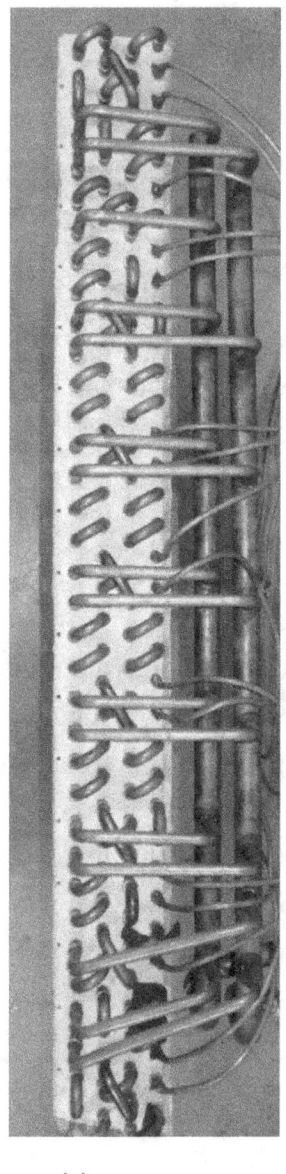 

     (a)   (b)

Figure 6.1 Prototypes of (a) Original Evaporator and (b) Optimized Evaporator

Operating the RTU with the new evaporator required the replacement of one other component. The new evaporator required different thermostatic expansion valves (TXVs). The first reason for this is that the optimized design includes stages that each have nine circuits whereas the original stages had eight. Each TXV needed to have nine circuit distributors in order to accommodate the new design. The second reason for the new TXVs is that the optimized design requires a larger refrigerant mass flow rate than the original design at the same level of superheat. The optimized design was generated with the same refrigerant exit state as the original design and, according to the simulation results, the TXVs must deliver 7.9 % more refrigerant flow at that condition in order to maintain this state with the optimized evaporator. For these reasons, TXVs with 9 circuit distributors and adjustable spring settings were installed on each stage of the RTU.

The optimized evaporator and new TXVs were installed in the RTU, and the unit was charged with refrigerant R410A. In order to match the operating conditions of the tests with those of the original unit, the flow rate and temperature of the water used to cool the condensers were adjusted along with the amount of refrigerant in the system and the spring settings on the TXVs. Best efforts attempted to match the liquid line pressure and temperature for each stage and the superheat at the exit from each evaporator header.

As stated in Chapter 3, the results of these performance measurements are needed to compare against the performance measurements with the original evaporator. Considering the magnitude of the expected improvement, it was necessary to measure the RTU performance with a low level of measurement uncertainty. For this reason, nine independent measurements of the system performance were acquired with the optimized evaporator over a two week period in the same manner as described in Chapter 3. The average value from these nine tests shows a total capacity of (27.01 ± 0.25) kW with the optimized evaporator and a system COP of 3.88 ± 0.04. Relevant portions of the recorded data are provided in Appendix B. The tabulated results are shown in Table 6.1.1.

Table 6.1.1 Refrigerant Side Capacity and COP Measurements for System with Optimized Evaporator

Test Number	Capacity stage 1 (kW)	Capacity stage 2 (kW)	Capacity total (kW)	Stage 1 Comp power (W)	COP Stage 1	Stage 2 Comp power (W)	COP Stage 2	System COP w/o blower	Blower Power (W)	System COP
1	13.28±0.14	13.70±0.15	26.98±0.21	2913±8	4.56±0.05	3048±8	4.50±0.05	4.53±0.03	991±9	3.88±0.03
2	13.23±0.16	13.61±0.23	26.84±0.28	2918±8	4.54±0.06	3040±8	4.48±0.08	4.50±0.05	986±11	3.87±0.04
3	13.29±0.15	13.75±0.17	27.04±0.23	2912±8	4.56±0.05	3048±8	4.51±0.06	4.54±0.04	999±16	3.89±0.03
4	13.26±0.15	13.71±0.15	26.97±0.21	2911±8	4.56±0.05	3041±8	4.51±0.05	4.53±0.04	983±9	3.89±0.03
5	13.26±0.15	13.62±0.16	26.88±0.22	2902±8	4.57±0.05	3045±8	4.47±0.05	4.52±0.04	1012±10	3.86±0.03
6	13.34±0.15	13.86±0.18	27.20±0.23	2906±8	4.59±0.05	3059±8	4.53±0.06	4.56±0.04	998±13	3.91±0.04
7	13.32±0.14	13.80±0.18	27.11±0.23	2902±8	4.59±0.05	3061±9	4.51±0.06	4.55±0.04	1003±15	3.89±0.03
8	13.27±0.15	13.73±0.19	26.99±0.24	2910±9	4.56±0.05	3042±9	4.51±0.06	4.54±0.04	1010±17	3.88±0.04
9	13.25±0.16	13.67±0.19	26.92±0.24	2919±9	4.54±0.06	3040±8	4.50±0.06	4.52±0.04	987±12	3.88±0.04
Average	13.28±0.15	13.72±0.18	27.01±0.25	2910±8	4.56±0.05	3047±8	4.50±0.06	4.53±0.04	997±13	3.88±0.04

The expressed values are shown as the average of all measurements ± the total combined uncertainty at 95% confidence as described in Appendix C.

6.2 Discussion of Measured Performance Improvement

The system capacity with the optimized evaporator was 2.2 % larger than the capacity with the original evaporator. The measured increase may not be large, but it is significant. As shown in Figure 6.2.1, the upper limit of the 95 % confidence interval for the system capacity with the original evaporator is marginally lower than the lower limit of the 95 % confidence interval for the system capacity with the optimized evaporator. However, it should be noted that the 2.2 % capacity increase is in line with the improvement prediction by the analytical method in Domanski (1988) which estimates a 2.9 % system capacity increase if the evaporator capacity were increased by 7.9 %.

It is interesting to examine the capacity for the individual stages. The Stage 1 capacity did not change significantly in response to the new evaporator but the Stage 2 capacity changed by more than 4 %. Figure 6.2.1 shows the capacity measured during both sets of tests; the individual stages are shown on the left and the total system capacity to the right.

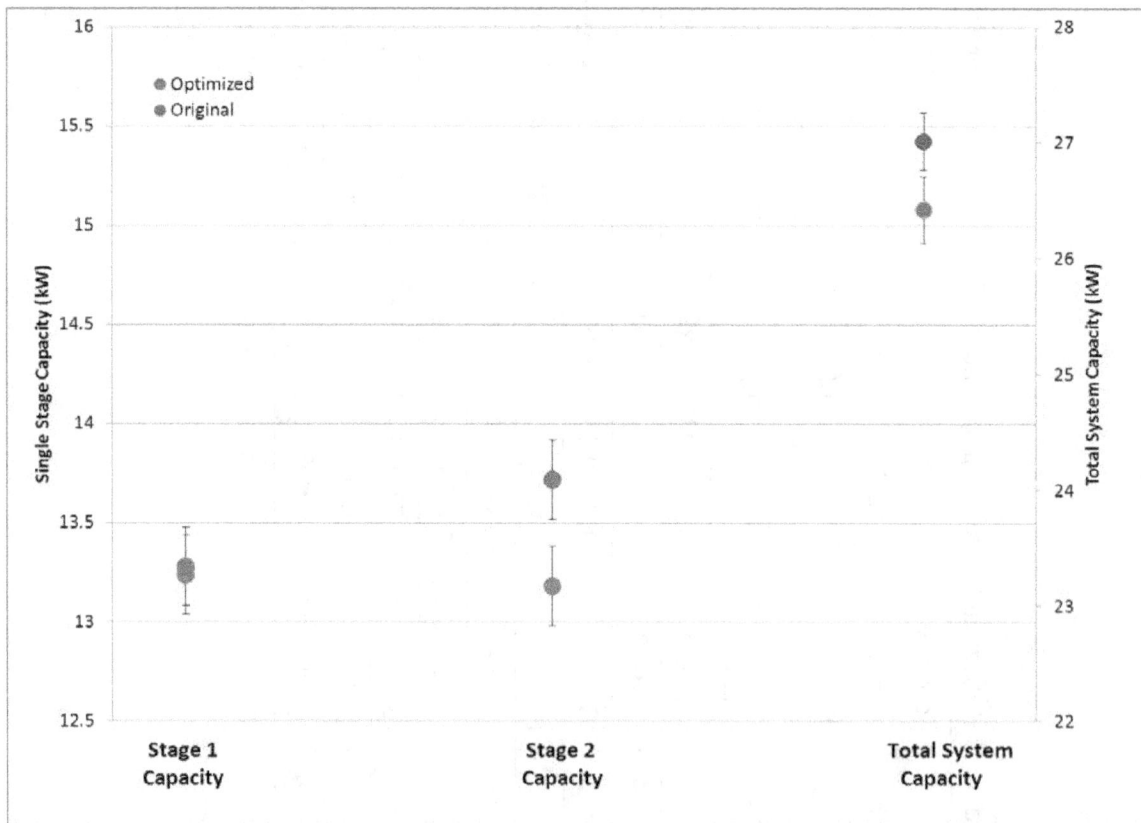

Figure 6.2.1 Measured Cooling Capacity with Original and Optimized Evaporator

The COP of the system with the optimized evaporator is 3.88±0.04, which represents 3.1 %
increase in efficiency over the system with the original evaporator.   Examining the
change in COP of the individual stages is helpful in order to determine where the
performance enhancement was realized.   Figure 6.2.2 compares the COP of each stage
and the system as measured during both sets of tests.   The data shows that the COP for
Stage 2 realized a significant increase of 4.7 %, while Stage 1 did not.   The efficiency of
the system as a whole did increase beyond the measurement uncertainty, and is shown
both with and without consideration of the blower power in Figure 6.2.2.

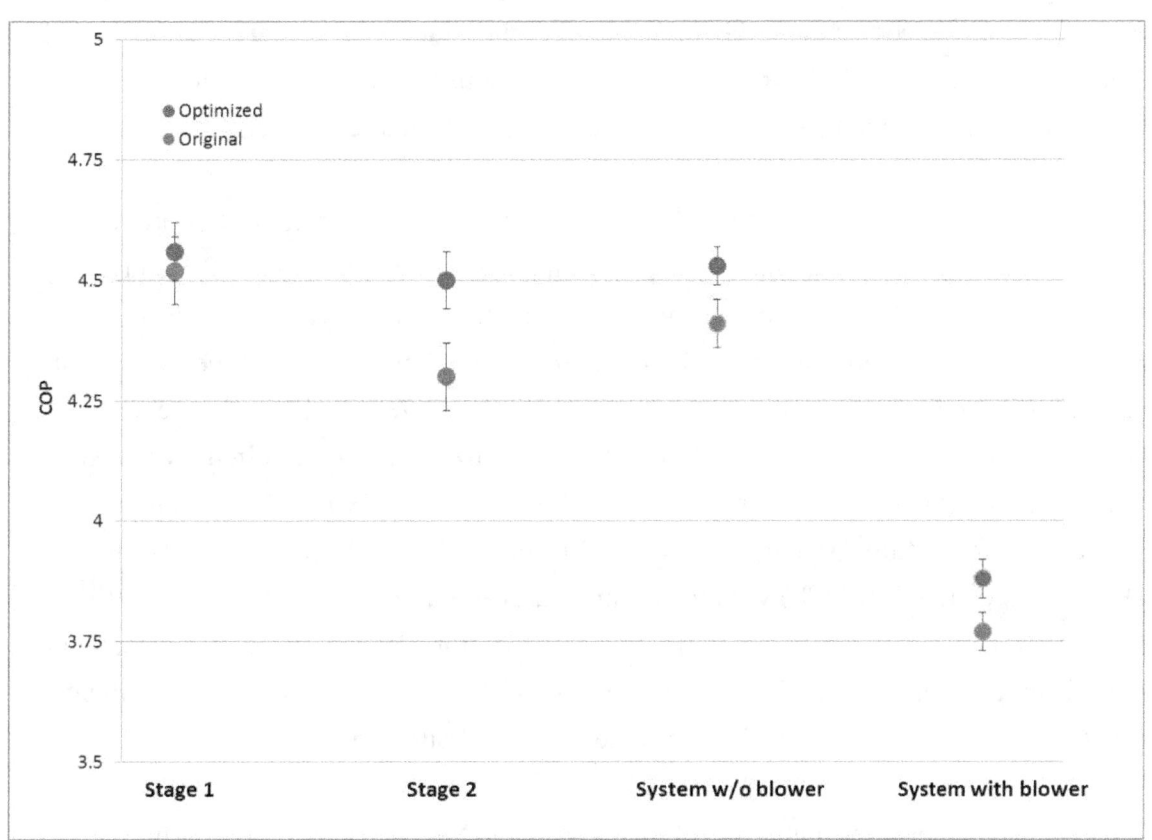

Figure 6.2.2 Measured COP for Stage 1, Stage 2, and System with Original and
Optimized Evaporator

Closer examination of the data revealed that the Stage 1 condenser subcooling was too
low during the tests with the optimized evaporator.   Tables 6.2.1 and 6.2.2 show data
extracted from Appendices A and B related to the RTU operation during both sets of tests.
The target conditions were intended to match the amount of superheat at the exit of each
evaporator stage and simultaneously match the liquid line saturation temperature and

subcooling. The data in Table 6.2.2 shows very good agreement in these parameters for Stage 2; however, Table 6.2.1 shows a (2.7±0.3) K ((5.0±0.5) °F) difference in the level of subcooling for Stage 1. This difference in subcooling effectively reduced the enthalpy difference between the evaporator inlet and exit by 3.5 %, which would result in a comparable reduction in cooling capacity. The fact that the optimized Stage 1 portion of the evaporator shows lower subcooling for the same liquid pressure and evaporator exit superheat indicates that there was not enough refrigerant in the Stage 1 loop during these tests. Had the Stage 1 loop been charged with the proper amount of refrigerant, it is likely that Stage 1 would have realized a performance enhancement comparable to that realized by Stage 2. It was difficult to accurately charge the unit with the exact amount of refrigerant needed to mimic the manufacturer's test data because the modifications that were made to the unit (replacing the air-cooled condensers with water-cooled brazed-plate heat exchangers) reduced the total required charge by nearly 70 %.

Nevertheless, the argument is still valid that the optimized evaporator did improve the performance of the system. One key aspect of the data shown in Tables 6.2.1 and 6.2.2 is that the exit evaporation temperature was significantly higher in each stage during the tests with the optimized evaporator than with the original evaporator. The evaporator saturation temperature for Stage 1 was (11.5±0.1) °C ((52.7±0.2) °F) under the test conditions with the original evaporator and the optimized evaporator circuitry raised it to (12.4±0.1) °C ((54.3±0.2) °F); an increase of (0.9±0.1) K ((1.7±0.2) °F). Similarly, the saturation temperature for Stage 2 was raised from (11.4±0.1) °C ((52.5±0.2) °F) to (12.1±0.1) °C ((53.8±0.2) °F) with the optimized evaporator, an increase of (0.7±0.1) K ((1.3±0.2) °F). This increase in evaporating temperature realized in each stage is quite significant compared to the limits of the 95 % confidence intervals and it does indicate that there is a real benefit resulting from the optimized circuitry.

This increase in evaporating temperature comes from two aspects of the design. First, this design provides a better match between the air and refrigerant temperature profiles; therefore, the temperature difference between the air and refrigerant is reduced. Second, since the optimized evaporator consists of the same amount of heat transfer area as the original evaporator, but provides 18 circuits instead of 16, the overall refrigerant pressure drop through the optimized evaporator is lower. Both of these differences contribute to an improvement in the overall RTU performance.

Overall, the data collected demonstrated that the optimized circuitry design does improve the efficiency of the system. Due to the magnitude of the capacity improvement and the slightly off-design charge amount, the data at first sight did not conclusively demonstrate the improvement beyond the uncertainty of the measured parameters. However, examination of the parameters pertaining to the individual stages does conclusively demonstrate that a real improvement exists. The benefit resulting from the optimized refrigerant circuitry in one of the two stages was offset by the influence of improper refrigerant charge. Had both stages been charged with the exact amount of refrigerant for the tests with the optimized evaporator, these tests would have enabled an exact apples-to-apples comparison which would have shown a more significant improvement as evidenced by the increased evaporating temperature in both stages.

Table 6.2.1 Comparison of Stage 1 Operation Between Tests with Original and Optimized Evaporators

	Liquid Line Parameters				Suction Line Parameters			
	Pressure (kPa)	Saturation Temperature (°C)	Temperature (°C)	Subcooling (K)	Pressure (kPa)	Saturation Temperature (°C)	Temperature (°C)	Superheat (K)
Original Evaporator	2695.0±6.3	44.4±0.1	37.8±0.2	6.6±0.2	1137.4±2.1	11.5±0.1	16.1±0.2	4.6±0.2
Optimized Evaporator	2696.9±6.4	44.4±0.1	40.6±0.2	3.9±0.2	1169.0±4.0	12.4±0.1	17.36±0.3	4.9±0.3

The expressed values are shown as the average of all measurements ± the total combined uncertainty at 95% confidence as described in Appendix C.

Table 6.2.2 Comparison of Stage 2 Operation Between Tests with Original and Optimized Evaporators

	Liquid Line Parameters				Suction Line Parameters			
	Pressure (kPa)	Saturation Temperature (°C)	Temperature (°C)	Subcooling (K)	Pressure (kPa)	Saturation Temperature (°C)	Temperature (°C)	Superheat (K)
Original Evaporator	2767.9±4.7	45.5±0.1	36.9±0.4	8.6±0.4	1133.2±2.1	11.4±0.1	15.6±0.2	4.2±0.2
Optimized Evaporator	2775.4±7.4	45.6±0.1	36.9±0.4	8.8±0.4	1157.6±5.7	12.1±0.1	16.3±0.2	4.1±0.3

The expressed values are shown as the average of all measurements ± the total combined uncertainty at 95% confidence as described in Appendix C.

7. SUMMARY

This study demonstrates the improvement in the 7.5 Ton RTU performance that was achieved by applying the evolutionary algorithms for optimizing the refrigerant circuitry of the evaporator.

First, the performance of the RTU was benchmarked by conducting performance tests in a conditioned indoor chamber. The unit was modified by replacing the air-cooled condensers with water-cooled heat exchangers in order to facilitate testing. The unit was operated in such a manner as to match the operational parameters for the evaporator section to those from the manufacturer's rating data. The results of the performance benchmark for the unit operating with original evaporator showed a capacity of (26.42 ± 0.29) kW and a COP of 3.77 ± 0.04.

Next, the in-situ air velocity profile through the evaporator was measured using Particle Image Velocimetry (PIV). The PIV measurements were recorded over several segments of the heat exchanger and were used to generate a map of the distribution of the air velocity passing through the heat exchanger.

A model of the evaporator was generated using the NIST simulation tool for finned-tube heat exchangers, EVAP-COND, which included the air velocity distribution characterized using the PIV data. The EVAP-COND model of the heat exchanger was tuned to the laboratory measurements, and the simulated performance resulted in a capacity of 26.38 kW. The Intelligent System for Heat Exchanger Design (ISHED), embedded in EVAP-COND, was implemented to improve the performance of the heat exchanger by redesigning the tube connection sequence. ISHED produced a design with 18 circuits and a simulated capacity of 28.46 kW, which is an increase of 7.9 %. Since ISHED only modifies the tube connection sequence and does not alter the heat exchanger layout, the size, shape, and material weight of the ISHED design are identical to the original design. A prototype of the optimized evaporator was produced by the original RTU manufacturer.

The original evaporator was replaced by the optimized prototype, and a new set of performance tests were carried out. The results of the tests showed that the RTU capacity changed from (26.42 ± 0.29) kW to (27.01 ± 0.25) kW, a 2.2 % increase. The data also showed that the COP increased from 3.77 ± 0.04 to 3.88 ± 0.04 with the new evaporator, an increase of 3.1 %.

The overall increase in cooling capacity did not provide a strong indicator of the performance enhancement because the lower limit on the 95 % confidence interval surrounding the capacity measurement with the optimized evaporator is very close to the upper confidence limit around the capacity measured with the original evaporator.

The measured improvement in the RTU performance was corroborated by a change in the evaporator saturation temperature. The optimized refrigerant circuitry caused an increase in the evaporating temperature in both stages, (0.9 ± 0.1) K $((1.7 \pm 0.2)$ °F$)$ for Stage 1 and (0.7 ± 0.1) K $((1.3 \pm 0.2)$ °F$)$ for Stage 2, which indicates that the performance improvement is real. However, closer examination of the data showed that the all of the improvement was realized in only one of the two stages, and it became apparent that the stage which did not realize any significant improvement was slightly undercharged with refrigerant during the tests with the optimized evaporator. This undercharge offset the optimized circuitry benefit that would have been realized by that stage.

8. REFERENCES

AMCA 2007.   ANSI/AMCA 210-07, *Laboratory Methods of Testing Fans for Certified Aerodynamic Performance Rating*. Air Movement and Control Association, Arlington Heights, IL

ASHRAE 2009. ANSI/ASHRAE Standard 37-2009, *Methods of Testing for Rating Electrically Driven Unitary Air-Conditioning and Heat Pump Equipment*. American Society of Heating, Refrigerating and Air-Conditioning Engineers, Atlanta, GA.

ARI 2007. Standard 340/360-2007, *Performance Rating of Commercial and Industrial Unitary Air-Conditioning and Heat Pump Equipment*. Air-Conditioning ,Heating, and Refrigeration Institute, Arlington, VA.

Chwalowski, M., Didion, D. A., Domanski, P. A. 1989. *Verification of Evaporator Computer Models and Analysis of Performance of an Evaporator Coil*. ASHRAE Transactions 95(1): 793-802.

Domanski, P.A. *Rating of Mixed Split Residential Air Conditioners*.   Proceedings of the Fifth Symposium on Improving Building Systems in Hot and Humid Climates, Houston, TX.   September 12-14, 1988.

Domanski, P.A., and Yashar, D.A., 2010.   *EVAP-COND Version 3-Simulation Models for Finned Tube Heat Exchangers with Circuitry Optimization*.   National Institute of Standards and Technology, Gaithersburg, MD.
http://www.nist.gov/el/building_environment/evapcond_software.cfm

Domanski, P.A., Yashar, D. A., and Lee, S.,   *Evaporator Optimization for Non-Uniform Air Distribution*.   Sustainable Refrigeration and Heat Pump Technology.   June 13-16, 2010, Stockholm, Sweden.

Fagan, T. J. 1980.   *The effects of air flow maldistributions on air-to-refrigerant heat exchanger performance*.   ASHRAE Transactions 86(2): 699-715.

Kaern, M.R., Elmegaard, B., and Larsen, L.F.S., 2013.*Comparison of fin-and-tube interlaced and face split evaporators with flow maldistribution and compensation*, Int. J. of Ref. 36: 203-214.

Lemmon, E.W., Huber, M.L., McLinden, M.O., 2010. *NIST Standard Reference Database 23: Reference Fluid Thermodynamic and Transport Properties-REFPROP, Version 9.0*, National Institute of Standards and Technology, Standard Reference Data Program, Gaithersburg.

Lemmon, E.W. and Jacobsen, R.T. *Equations of State for Mixtures of R-32, R-125, R-134a, and R152a*. J. Phys. Chem. Ref. Data, 33(2), 2004.

Payne, W. V. and Domanski, P. A. *Potential Benefits of Smart Refrigerant Distributors* (ARTI-21CR/605-200-50-01; 196 p.) January 01, 2003.

Taylor, B.N., and Kuyatt, C.E., 1994. *Guidelines for evaluating and expressing the uncertainty of NIST measurement results*, NIST Technical Note 1297, 1994 edition, U.S. Department of Commerce.

Yashar, D.A., Wojtusiak, J., Kaufman, K., and Domanski, P.A., 2012. *A Dual Mode Evolutionary Algorithm for Designing Optimized Refrigerant Circuitries for Finned-Tube Heat Exchangers* HVAC&R Research 18(5), SPECIAL ISSUE: Optimization in HVAC&R, 834-844.

Yashar, D.A., and Cho, H.H., 2007. *Air-Side Velocity Distribution in Finned-Tube Heat Exchangers*. NIST Internal Report 7474, U.S. Department of Commerce, National Institute of Standards and Technology, Gaithersburg, MD.

Appendix A – Data recorded during performance measurement tests with original heat exchanger

Table A.1 - Test 1 with Original Evaporator

Scan	Stage 1						Stage 2						
	Evaporator Outlet Temp (°C)	Evaporator Outlet Pressure (kPa)	Liquid Line Temp (°C)	Liquid Line Pressure (kPa)	Mass Flow Rate (gs⁻¹)	Compressor Power (W)	Evaporator Outlet Temp (°C)	Evaporator Outlet Pressure (kPa)	Liquid Line Temp (°C)	Liquid Line Pressure (kPa)	Mass Flow Rate (gs⁻¹)	Compressor Power (W)	Blower Power (W)
1	15.823	1139.35	37.689	2692.28	0.0805	2927	15.398	1130.83	37.345	2763.66	0.0775	3072	1058
2	15.960	1139.68	37.795	2695.24	0.0786	2939	15.392	1130.01	37.505	2764.65	0.0771	3072	1046
3	15.989	1138.76	37.421	2700.74	0.0783	2932	15.383	1130.44	37.196	2769.76	0.0791	3075	995
4	15.939	1137.04	37.676	2690.49	0.0799	2920	15.378	1130.77	37.270	2762.03	0.0788	3067	1010
5	15.953	1138.34	37.456	2688.37	0.0796	2925	15.372	1129.61	37.318	2761.28	0.0798	3070	1049
6	15.976	1137.83	37.514	2689.77	0.0710	2920	15.419	1131.41	37.259	2767.28	0.0798	3060	1061
Avg	15.940	1138.50	37.592	2692.81	0.0795	2927.2	15.390	1130.51	37.316	2764.78	0.0787	3069.3	1036.5
StDev	0.060	0.98	0.1493	4.55	0.00092	7.36	0.0172	0.64	0.1063	3.23	0.00114	5.28	27.33

Table A.2 - Test 2 with Original Evaporator

Scan	Stage 1						Stage 2						
	Evaporator Outlet Temp (°C)	Evaporator Outlet Pressure (kPa)	Liquid Line Temp (°C)	Liquid Line Pressure (kPa)	Mass Flow Rate (gs⁻¹)	Compressor Power (W)	Evaporator Outlet Temp (°C)	Evaporator Outlet Pressure (kPa)	Liquid Line Temp (°C)	Liquid Line Pressure (kPa)	Mass Flow Rate (gs⁻¹)	Compressor Power (W)	Blower Power (W)
1	15.846	1137.07	37.800	2693.12	0.07824	2925	15.433	1132.07	37.432	2763.00	0.07994	3069	984
2	15.924	1126.32	37.766	2685.17	0.07930	2919	15.389	1130.50	37.507	2764.11	0.07924	3070	1047
3	16.160	1136.09	37.319	2686.56	0.07864	2924	15.404	1130.33	37.261	2762.65	0.07707	3069	948
4	15.845	1135.76	37.491	2687.79	0.07881	2920	15.415	1131.06	37.295	2768.99	0.07876	3062	1000
5	15.957	1136.56	37.797	2689.35	0.07837	2921	15.422	1132.54	37.541	2763.41	0.07982	3073	1000
6	15.852	1135.87	37.790	2687.55	0.08051	2917	15.426	1133.58	37.239	2768.40	0.07981	3061	1007
Avg	15.931	1134.61	37.660	2688.26	0.07898	2921	15.415	1131.68	37.379	2765.09	0.07911	3067.3	997.7
StDev	0.122	4.09	0.205	2.76	0.00084	3.03	0.016	1.27	0.131	2.84	0.00109	4.76	32.20

Table A.3 - Test 3 with Original Evaporator

Scan	Stage 1						Stage 2						
	Evaporator Outlet Temp (°C)	Evaporator Outlet Pressure (kPa)	Liquid Line Temp (°C)	Liquid Line Pressure (kPa)	Mass Flow Rate (gs⁻¹)	Compressor Power (W)	Evaporator Outlet Temp (°C)	Evaporator Outlet Pressure (kPa)	Liquid Line Temp (°C)	Liquid Line Pressure (kPa)	Mass Flow Rate (gs⁻¹)	Compressor Power (W)	Blower Power (W)
1	15.963	1137.93	37.678	2692.76	0.07823	2927	15.454	1132.89	37.124	2762.12	0.07994	3067	1000
2	15.987	1136.92	37.575	2689.24	0.08054	2916	15.470	1132.34	37.100	2777.31	0.07924	3060	996
3	15.962	1137.62	37.748	2692.12	0.07921	2921	15.476	1129.73	37.212	2760.30	0.07894	3065	984
4	15.934	1138.91	37.712	2695.21	0.07897	2939	15.446	1130.36	37.104	2772.66	0.07924	3077	1030
5	15.877	1139.83	37.650	2686.84	0.08037	2920	15.325	1130.08	36.850	2764.44	0.07728	3062	1012
6	15.752	1138.19	37.562	2696.19	0.07972	2920	15.361	1131.47	37.031	2767.72	0.07721	3081	1035
Avg	15.912	1138.23	37.654	2692.06	0.07951	2923.8	15.422	1131.14	37.070	2767.43	0.07864	3068.7	1009.5
StDev	0.092	1.02	0.074	3.54	0.00088	8.23	0.064	1.29	0.122	6.53	0.00113	8.45	20.00

Table A.4 - Test 4 with Original Evaporator

Scan	Stage 1						Stage 2						
	Evaporator Outlet Temp (°C)	Evaporator Outlet Pressure (kPa)	Liquid Line Temp (°C)	Liquid Line Pressure (kPa)	Mass Flow Rate (gs⁻¹)	Compressor Power (W)	Evaporator Outlet Temp (°C)	Evaporator Outlet Pressure (kPa)	Liquid Line Temp (°C)	Liquid Line Pressure (kPa)	Mass Flow Rate (gs⁻¹)	Compressor Power (W)	Blower Power (W)
1	15.805	1139.07	37.828	2699.31	0.07958	2935	15.390	1132.35	37.144	2765.44	0.07773	3070	1046
2	15.825	1136.36	37.679	2700.03	0.07947	2937	15.376	1131.73	37.002	2771.02	0.07876	3068	987
3	15.872	1135.61	37.745	2699.72	0.07908	2939	15.481	1133.13	37.016	2770.29	0.07982	3068	1010
4	15.837	1136.23	37.495	2700.17	0.08047	2933	15.482	1133.31	37.155	2767.37	0.07894	3072	1042
5	15.886	1132.47	38.016	2698.25	0.07886	2930	15.517	1133.41	37.048	2772.58	0.07923	3073	988
6	15.811	1138.25	37.959	2700.30	0.07946	2935	15.528	1134.86	37.280	2770.03	0.07798	3083	1043
Avg	15.839	1136.33	37.787	2699.63	0.07949	2934.8	15.462	1133.13	37.108	2769.45	0.07874	3072.3	1019.3
StDev	0.034	2.31	0.191	0.76	0.00055	3.13	0.065	1.06	0.106	2.59	0.00078	5.61	27.93

Table A.5 - Test 5 with Original Evaporator

Scan	Stage 1						Stage 2						
	Evaporator Outlet Temp (°C)	Evaporator Outlet Pressure (kPa)	Liquid Line Temp (°C)	Liquid Line Pressure (kPa)	Mass Flow Rate (gs⁻¹)	Compressor Power (W)	Evaporator Outlet Temp (°C)	Evaporator Outlet Pressure (kPa)	Liquid Line Temp (°C)	Liquid Line Pressure (kPa)	Mass Flow Rate (gs⁻¹)	Compressor Power (W)	Blower Power (W)
1	15.927	1138.91	37.644	2687.30	0.07833	2922	15.616	1133.55	37.111	2772.00	0.07821	3065	1002
2	15.779	1139.83	37.916	2694.77	0.07818	2926	15.643	1135.30	37.156	2775.25	0.07923	3061	1038
3	16.204	1137.21	37.872	2696.47	0.07962	2927	15.690	1135.77	37.184	2767.39	0.07728	3068	1050
4	16.264	1137.08	37.714	2699.39	0.07929	2935	15.683	1135.50	36.985	2768.54	0.07769	3068	998
5	16.194	1135.63	37.992	2700.16	0.07835	2924	15.734	1136.46	37.118	2769.14	0.07713	3070	1058
6	16.220	1136.08	37.852	2694.09	0.07804	2923	15.726	1136.29	36.867	2775.54	0.07876	3067	1035
Avg	16.098	1137.46	37.832	2695.36	0.07864	2926.2	15.682	1135.48	37.070	2771.31	0.07805	3066.5	1030.2
StDev	0.197	1.63	0.129	4.64	0.00065	4.71	0.046	1.05	0.121	3.51	0.00084	3.15	24.82

Table A.6 - Test 6 with Original Evaporator

Scan	Stage 1						Stage 2						
	Evaporator Outlet Temp (°C)	Evaporator Outlet Pressure (kPa)	Liquid Line Temp (°C)	Liquid Line Pressure (kPa)	Mass Flow Rate (gs⁻¹)	Compressor Power (W)	Evaporator Outlet Temp (°C)	Evaporator Outlet Pressure (kPa)	Liquid Line Temp (°C)	Liquid Line Pressure (kPa)	Mass Flow Rate (gs⁻¹)	Compressor Power (W)	Blower Power (W)
1	16.272	1137.36	37.897	2707.35	0.07961	2934	15.745	1136.26	36.808	2776.85	0.07765	3065	997
2	16.215	1139.33	38.254	2716.48	0.07996	2943	15.805	1135.61	37.146	2768.88	0.07906	3073	1015
3	16.309	1137.76	38.132	2699.41	0.07828	2931	15.790	1134.19	36.987	2761.89	0.07976	3073	1036
4	16.218	1137.79	37.933	2695.89	0.08054	2931	15.786	1135.54	36.750	2768.45	0.07982	3072	1060
5	16.267	1138.56	37.466	2697.49	0.07829	2927	15.801	1135.38	36.698	2770.60	0.07982	3073	1047
6	16.290	1138.16	37.638	2692.89	0.08164	2925	15.788	1135.30	36.547	2778.40	0.07894	3069	981
Avg	16.262	1138.16	37.887	2701.59	0.07972	2931.8	15.786	1135.38	36.823	2770.84	0.07917	3070.8	1022.7
StDev	0.039	0.703	0.295	8.77	0.00131	6.34	0.022	0.675	0.214	6.05	0.00084	3.25	30.39

Table A.7 - Test 7 with Original Evaporator

Scan	Stage 1						Stage 2						
	Evaporator Outlet Temp (°C)	Evaporator Outlet Pressure (kPa)	Liquid Line Temp (°C)	Liquid Line Pressure (kPa)	Mass Flow Rate (gs⁻¹)	Compressor Power (W)	Evaporator Outlet Temp (°C)	Evaporator Outlet Pressure (kPa)	Liquid Line Temp (°C)	Liquid Line Pressure (kPa)	Mass Flow Rate (gs⁻¹)	Compressor Power (W)	Blower Power (W)
1	16.295	1138.08	37.812	2696.77	0.08068	2929	15.827	1134.80	36.589	2771.45	0.07924	3069	986
2	16.336	1139.42	37.815	2705.49	0.08039	2932	15.814	1134.03	36.592	2775.03	0.07780	3071	1011
3	16.421	1136.74	37.943	2700.24	0.07942	2933	15.813	1133.68	36.556	2762.66	0.07721	3053	1023
4	16.336	1137.12	37.748	2699.68	0.08041	2933	15.878	1133.13	36.234	2761.41	0.07773	3047	1042
5	16.335	1138.26	37.724	2697.88	0.08058	2932	15.896	1134.03	36.100	2769.20	0.07876	3040	1054
6	16.359	1138.41	38.072	2699.21	0.07916	2929	15.873	1133.70	36.247	2767.96	0.07957	3042	991
Avg	16.347	1138.00	37.852	2699.88	0.08011	2931.3	15.850	1133.90	36.386	2767.95	0.07838	3053.7	1017.8
StDev	0.043	0.963	0.132	3.02	0.00065	1.86	0.037	0.550	0.218	5.19	0.00094	13.44	27.20

Table A.8 - Test 8 with Original Evaporator

Scan	Stage 1						Stage 2						
	Evaporator Outlet Temp (°C)	Evaporator Outlet Pressure (kPa)	Liquid Line Temp (°C)	Liquid Line Pressure (kPa)	Mass Flow Rate (gs⁻¹)	Compressor Power (W)	Evaporator Outlet Temp (°C)	Evaporator Outlet Pressure (kPa)	Liquid Line Temp (°C)	Liquid Line Pressure (kPa)	Mass Flow Rate (gs⁻¹)	Compressor Power (W)	Blower Power (W)
1	16.253	1137.72	37.920	2693.36	0.07989	2923	15.860	1135.40	36.091	2769.01	0.07899	3053	984
2	16.315	1137.45	37.778	2681.93	0.07907	2925	15.832	1135.99	36.284	2768.78	0.07715	3051	991
3	16.366	1137.89	37.882	2689.61	0.07908	2924	15.868	1133.53	36.448	2762.52	0.07731	3048	1011
4	16.229	1137.18	37.964	2689.08	0.07979	2921	15.838	1134.56	36.374	2761.12	0.07899	3044	1028
5	16.170	1136.83	37.685	2694.76	0.07941	2932	15.839	1132.84	36.403	2767.86	0.07907	3056	978
6	15.922	1137.76	37.868	2694.57	0.07888	2925	15.829	1133.74	36.842	2767.52	0.07888	3048	987
Avg	16.209	1137.47	37.849	2690.55	0.07935	2925	15.844	1134.34	36.407	2766.13	0.07826	3050	996.5
StDev	0.158	0.407	0.102	4.88	0.00042	3.74	0.017	1.198	0.248	3.42	0.00088	4.24	19.09

Appendix B – Data recorded during performance measurement tests with optimized heat exchanger

Table B.1 - Test 1 with Optimized Evaporator

Scan	Stage 1						Stage 2						
	Evaporator Outlet Temp (°C)	Evaporator Outlet Pressure (kPa)	Liquid Line Temp (°C)	Liquid Line Pressure (kPa)	Mass Flow Rate (gs⁻¹)	Compressor Power (W)	Evaporator Outlet Temp (°C)	Evaporator Outlet Pressure (kPa)	Liquid Line Temp (°C)	Liquid Line Pressure (kPa)	Mass Flow Rate (gs⁻¹)	Compressor Power (W)	Blower Power (W)
1	17.602	1170.08	40.709	2703.74	0.0812	2915	16.514	1159.06	36.917	2769.36	0.0812	3046	933
2	17.570	1170.17	40.735	2701.39	0.0820	2912	16.423	1158.84	36.783	2768.67	0.0817	3050	1010
3	17.752	1171.55	40.615	2694.47	0.0823	2908	16.541	1159.56	37.012	2771.99	0.0841	3048	1022
4	17.602	1171.66	40.542	2691.21	0.0816	2906	16.411	1160.20	36.932	2772.67	0.0805	3046	987
5	17.614	1172.30	40.634	2704.72	0.0806	2917	16.448	1160.14	37.043	2770.70	0.0816	3048	979
6	17.556	1171.24	40.709	2705.02	0.0812	2920	16.479	1161.58	36.829	2771.88	0.0806	3049	1017
Avg	17.616	1171.17	40.657	2700.09	0.0815	2913.0	16.469	1159.90	36.920	2770.88	0.0816	3047.8	991.3
StDev	0.071	0.88	0.073	5.85	0.00063	5.37	0.052	0.99	0.101	1.59	0.00131	1.60	33.23

Table B.2 - Test 2 with Optimized Evaporator

Scan	Stage 1						Stage 2						
	Evaporator Outlet Temp (°C)	Evaporator Outlet Pressure (kPa)	Liquid Line Temp (°C)	Liquid Line Pressure (kPa)	Mass Flow Rate (gs⁻¹)	Compressor Power (W)	Evaporator Outlet Temp (°C)	Evaporator Outlet Pressure (kPa)	Liquid Line Temp (°C)	Liquid Line Pressure (kPa)	Mass Flow Rate (gs⁻¹)	Compressor Power (W)	Blower Power (W)
1	17.103	1173.51	40.784	2707.10	0.0816	2926	16.112	1165.84	37.487	2768.58	0.0821	3040	1029
2	17.240	1173.62	40.758	2693.73	0.0823	2912	16.022	1164.89	37.190	2770.97	0.0800	3040	1007
3	17.155	1174.07	40.639	2698.92	0.0811	2916	16.003	1164.54	37.392	2767.76	0.0813	3036	974
4	16.961	1171.51	40.718	2705.49	0.0806	2922	15.947	1163.20	37.249	2768.33	0.0853	3041	965
5	16.979	1171.37	40.676	2705.31	0.0804	2923	15.839	1163.23	37.343	2772.60	0.0810	3044	973
6	17.114	1169.36	40.742	2696.25	0.0819	2909	15.876	1163.24	37.485	2771.89	0.0799	3039	965
Avg	17.092	1172.24	40.720	2701.13	0.0813	2918.0	15.966	1164.16	37.358	2770.02	0.0816	3040.0	985.5
StDev	0.107	1.81	0.054	5.58	0.00072	6.72	0.101	1.11	0.122	2.06	0.00199	2.61	26.40

Table B.3 - Test 3 with Optimized Evaporator

Scan	Stage 1						Stage 2						Blower Power (W)
	Evaporator Outlet Temp (°C)	Evaporator Outlet Pressure (kPa)	Liquid Line Temp (°C)	Liquid Line Pressure (kPa)	Mass Flow Rate (gs⁻¹)	Compressor Power (W)	Evaporator Outlet Temp (°C)	Evaporator Outlet Pressure (kPa)	Liquid Line Temp (°C)	Liquid Line Pressure (kPa)	Mass Flow Rate (gs⁻¹)	Compressor Power (W)	
1	17.570	1170.17	40.735	2701.39	0.0820	2912	16.423	1158.84	36.783	2779.34	0.0817	3050	1010
2	17.752	1171.55	40.615	2694.47	0.0823	2908	16.541	1159.56	37.012	2779.27	0.0841	3048	1022
3	17.602	1171.66	40.542	2691.21	0.0816	2906	16.411	1160.20	36.932	2781.08	0.0805	3046	987
4	17.614	1172.30	40.634	2704.72	0.0806	2917	16.448	1160.14	37.043	2780.06	0.0816	3048	979
5	17.556	1171.24	40.709	2705.02	0.0812	2920	16.479	1161.58	36.829	2779.20	0.0806	3049	1017
6	17.587	1172.85	40.588	2689.97	0.0816	2907	16.422	1159.80	36.920	2777.64	0.0816	3049	979
Avg	17.613	1171.63	40.637	2697.80	0.0816	2911.7	16.454	1160.02	36.920	2779.43	0.0817	3048.3	999.0
StDev	0.072	0.92	0.073	6.76	0.00062	5.75	0.050	0.91	0.101	1.13	0.00130	1.37	19.59

Table B.4 - Test 4 with Optimized Evaporator

Scan	Stage 1						Stage 2						Blower Power (W)
	Evaporator Outlet Temp (°C)	Evaporator Outlet Pressure (kPa)	Liquid Line Temp (°C)	Liquid Line Pressure (kPa)	Mass Flow Rate (gs⁻¹)	Compressor Power (W)	Evaporator Outlet Temp (°C)	Evaporator Outlet Pressure (kPa)	Liquid Line Temp (°C)	Liquid Line Pressure (kPa)	Mass Flow Rate (gs⁻¹)	Compressor Power (W)	
1	17.311	1167.29	40.759	2700.59	0.0818	2917	16.268	1157.60	37.123	2771.51	0.0811	3042	1013
2	17.192	1170.39	40.464	2696.44	0.0803	2910	16.171	1157.96	37.108	2772.45	0.0814	3043	984
3	17.162	1168.56	40.622	2704.08	0.0806	2916	16.126	1157.52	36.845	2773.36	0.0824	3041	972
4	17.158	1167.68	40.651	2697.27	0.0822	2911	16.109	1157.70	36.877	2769.79	0.0807	3041	976
5	17.184	1167.15	40.511	2682.87	0.0818	2900	16.063	1156.72	36.938	2770.14	0.0818	3041	976
6	17.129	1165.14	40.649	2700.86	0.0815	2911	16.094	1155.90	36.953	2773.09	0.0811	3040	975
Avg	17.189	1167.70	40.609	2697.02	0.0813	2910.8	16.138	1157.23	36.974	2771.72	0.0814	3041.3	982.7
StDev	0.064	1.73	0.106	7.46	0.00074	6.05	0.073	0.77	0.117	1.51	0.00062	1.03	15.38

Table B.5 - Test 5 with Optimized Evaporator

	Stage 1						Stage 2						
Scan	Evaporator Outlet Temp (°C)	Evaporator Outlet Pressure (kPa)	Liquid Line Temp (°C)	Liquid Line Pressure (kPa)	Mass Flow Rate (gs^{-1})	Compressor Power (W)	Evaporator Outlet Temp (°C)	Evaporator Outlet Pressure (kPa)	Liquid Line Temp (°C)	Liquid Line Pressure (kPa)	Mass Flow Rate (gs^{-1})	Compressor Power (W)	Blower Power (W)
1	17.219	1161.73	40.279	2679.76	0.0817	2888	16.689	1145.00	35.839	2776.61	0.0798	3043	965
2	17.111	1162.86	40.348	2695.68	0.0799	2905	16.677	1144.54	35.775	2774.32	0.0818	3045	1011
3	17.299	1161.60	40.510	2699.42	0.0800	2909	16.638	1146.15	36.205	2776.14	0.0794	3044	1031
4	17.239	1161.80	40.611	2699.40	0.0814	2908	16.557	1148.74	36.471	2774.00	0.0801	3044	1023
5	17.326	1162.85	40.608	2692.05	0.0819	2905	16.499	1149.37	36.524	2774.54	0.0802	3046	1018
6	17.412	1165.09	40.430	2686.19	0.0826	2897	16.534	1151.81	36.332	2776.72	0.0812	3047	1021
Avg	17.268	1162.66	40.464	2692.08	0.0812	2902.0	16.599	1147.60	36.191	2775.39	0.0805	3044.8	1011.5
StDev	0.103	1.32	0.137	7.83	0.00108	8.05	0.080	2.84	0.318	1.23	0.00091	1.47	23.70

Table B.6 - Test 6 with Optimized Evaporator

	Stage 1						Stage 2						
Scan	Evaporator Outlet Temp (°C)	Evaporator Outlet Pressure (kPa)	Liquid Line Temp (°C)	Liquid Line Pressure (kPa)	Mass Flow Rate (gs^{-1})	Compressor Power (W)	Evaporator Outlet Temp (°C)	Evaporator Outlet Pressure (kPa)	Liquid Line Temp (°C)	Liquid Line Pressure (kPa)	Mass Flow Rate (gs^{-1})	Compressor Power (W)	Blower Power (W)
1	17.380	1170.18	40.269	2690.99	0.0819	2904	16.218	1159.41	36.974	2787.19	0.0838	3061	977
2	17.434	1171.65	40.283	2692.39	0.0814	2904	16.207	1159.20	37.148	2788.39	0.0817	3062	1012
3	17.451	1171.19	40.306	2700.26	0.0806	2909	16.230	1159.43	37.062	2783.40	0.0833	3062	1030
4	17.119	1168.91	40.409	2698.25	0.0811	2914	16.001	1160.83	37.082	2761.56	0.0830	3041	995
5	16.895	1169.79	40.592	2694.62	0.0823	2908	15.981	1158.30	37.109	2795.62	0.0831	3069	976
6	17.070	1170.98	40.446	2683.71	0.0825	2897	16.302	1159.87	36.914	2781.53	0.0849	3059	995
Avg	17.225	1170.45	40.384	2693.37	0.0816	2906.0	16.156	1159.51	37.048	2782.95	0.0833	3059.0	997.5
StDev	0.229	1.02	0.124	5.88	0.00073	5.76	0.134	0.83	0.088	11.56	0.00104	9.44	20.79

Table B.7 - Test 7 with Optimized Evaporator

Scan	Stage 1						Stage 2						
	Evaporator Outlet Temp (°C)	Evaporator Outlet Pressure (kPa)	Liquid Line Temp (°C)	Liquid Line Pressure (kPa)	Mass Flow Rate (gs⁻¹)	Compressor Power (W)	Evaporator Outlet Temp (°C)	Evaporator Outlet Pressure (kPa)	Liquid Line Temp (°C)	Liquid Line Pressure (kPa)	Mass Flow Rate (gs⁻¹)	Compressor Power (W)	Blower Power (W)
1	17.381	1165.40	40.351	2695.04	0.0811	2903	16.066	1152.69	36.901	2793.06	0.0832	3066	1016
2	17.420	1166.42	40.297	2697.25	0.0811	2903	16.116	1153.82	36.779	2791.45	0.0811	3064	1003
3	17.347	1165.82	40.285	2690.43	0.0815	2902	16.162	1153.72	36.778	2783.90	0.0816	3058	978
4	17.412	1165.89	40.314	2692.36	0.0815	2901	16.038	1153.95	36.631	2786.22	0.0812	3059	1032
5	17.293	1166.60	40.379	2691.25	0.0807	2904	16.016	1153.73	36.719	2785.28	0.0818	3064	977
6	17.317	1165.74	40.355	2691.91	0.0815	2900	16.001	1155.48	36.691	2783.83	0.0819	3055	1009
Avg	17.362	1165.98	40.330	2693.04	0.0813	2902.2	16.067	1153.90	36.750	2787.29	0.0818	3061.0	1002.5
StDev	0.052	0.45	0.037	2.59	0.00034	1.47	0.062	0.90	0.093	3.98	0.00074	4.29	21.66

Table B.8 - Test 8 with Optimized Evaporator

Scan	Stage 1						Stage 2						
	Evaporator Outlet Temp (°C)	Evaporator Outlet Pressure (kPa)	Liquid Line Temp (°C)	Liquid Line Pressure (kPa)	Mass Flow Rate (gs⁻¹)	Compressor Power (W)	Evaporator Outlet Temp (°C)	Evaporator Outlet Pressure (kPa)	Liquid Line Temp (°C)	Liquid Line Pressure (kPa)	Mass Flow Rate (gs⁻¹)	Compressor Power (W)	Blower Power (W)
1	17.492	1160.55	40.485	2694.69	0.0803	2906	16.042	1147.45	36.428	2765.08	0.0793	3037	1003
2	17.605	1161.89	40.599	2700.52	0.0806	2909	16.203	1149.28	36.615	2766.58	0.0808	3039	1021
3	17.617	1163.49	40.644	2698.72	0.0812	2914	16.264	1151.51	36.503	2770.74	0.0804	3040	1019
4	17.798	1166.05	40.585	2697.16	0.0812	2910	16.372	1152.92	36.484	2772.35	0.0803	3044	976
5	18.057	1168.67	40.605	2700.80	0.0812	2912	16.602	1152.61	36.752	2776.72	0.0816	3045	1031
6	18.071	1169.53	40.573	2698.88	0.0812	2912	16.685	1155.17	36.256	2772.89	0.0794	3044	1011
Avg	17.773	1165.03	40.582	2698.46	0.0809	2910.5	16.361	1151.49	36.506	2770.72	0.0803	3041.5	1010.2
StDev	0.245	3.66	0.053	2.28	0.00040	2.81	0.245	2.76	0.168	4.30	0.00085	3.27	19.23

Table B.9 - Test 9 with Optimized Evaporator

Scan	Stage 1						Stage 2						Blower Power (W)
	Evaporator Outlet Temp (°C)	Evaporator Outlet Pressure (kPa)	Liquid Line Temp (°C)	Liquid Line Pressure (kPa)	Mass Flow Rate (gs^{-1})	Compressor Power (W)	Evaporator Outlet Temp (°C)	Evaporator Outlet Pressure (kPa)	Liquid Line Temp (°C)	Liquid Line Pressure (kPa)	Mass Flow Rate (gs^{-1})	Compressor Power (W)	
1	17.291	1175.57	40.724	2690.38	0.0826	2915	16.177	1166.01	37.445	2769.46	0.0820	3045	963
2	17.213	1176.72	40.634	2705.62	0.0806	2925	16.092	1165.37	37.398	2770.40	0.0803	3036	985
3	17.103	1173.51	40.784	2707.10	0.0816	2926	16.112	1165.84	37.487	2768.58	0.0821	3040	1029
4	17.240	1173.62	40.758	2693.73	0.0823	2912	16.022	1164.89	37.190	2770.97	0.0800	3040	1007
5	17.155	1174.07	40.639	2698.92	0.0811	2916	16.003	1164.54	37.392	2767.76	0.0813	3036	974
6	16.961	1171.51	40.718	2705.49	0.0806	2922	15.947	1163.20	37.249	2768.33	0.0853	3041	965
Avg	17.161	1174.17	40.710	2700.21	0.0814	2919.3	16.059	1164.98	37.360	2769.25	0.0818	3039.7	987.2
StDev	0.118	1.81	0.061	7.00	0.00084	5.79	0.084	1.03	0.116	1.25	0.00190	3.39	26.06

Appendix C – Uncertainty analysis

The methods of calculating the reported values listed in this report are presented in this appendix. Each of the presented uncertainty values is derived from the methods outlined in Taylor and Kuyatt (1994) and is derived from two components. The first component, called type A uncertainty, is based on the variation of repeated measurement results and the second component, type B, is based on scientific judgment. The overall uncertainty is derived from a Pythagorean relationship between the two types of uncertainty.

The calculations presented in this Appendix use the data set from the first test of the RTU with the optimized heat exchanger. The same methods were applied to all data sets. Values for the type A uncertainty were calculated for each of the measured values as follows, based on a sampling size of six independent measurements; the data used to calculate the performance included in this report were based on many more scans, but the six scans reported for this data set in Appendix B are used in this analysis for illustrative purposes.

Parameter: Evaporator Outlet Temperature, stage 1
Average of 6 measurements: 17.616 °C
Standard deviation of the 6 measurements: 0.071 °C

First, the Standard Error surrounding this value is calculated by dividing the standard deviation by the square root of the number of samples

$$SE = \frac{0.071\ °C}{\sqrt{6}} = 0.0290\ °C$$

The Student t-value for two sided confidence limits at 95 % confidence based on 6 degrees of freedom is

$$t(0.975, 6) = 2.447$$

Therefore the bounds for the upper and lower confidence limits around the reported value are calculated by multiplying the t-value by the Standard Error

Confidence interval = 2.447 ∗ (0.0290 °C) = 0.071 °C

The type B uncertainty for this measurement is based on the calibration data for the instrument used to measure the parameter. In this case, the data was collected with a T-type thermocouple calibrated to with 0.15 °C of true value at 95 % confidence.

Therefore the total uncertainty for the stage 1 evaporator outlet temperature is calculated by the square root of the sum of the squares of the type A and type B uncertainties

$$U = \sqrt{(0.071\ °C)^2 + (0.150\ °C)^2} = 0.166\ °C$$

These calculations were repeated for each of the measured parameters listed in the data set in Appendix B and the values are shown in the table below.

Table C.1 – Calculation of Total Uncertainty for Measured Parameters, Test 1 with Optimized Evaporator

	Parameter	Average of Measurements	Standard Deviation	Standard Error	Confidence Interval	Measurement bias	Total Uncertainty
Stage 1	Mass Flow Rate (g/s)	0.08149	0.00063	0.00026	0.00063	0.00081	0.00103
	Evaporator Outlet Pressure (kPa)	1171.17	0.88	0.36	0.88	3.50	3.61
	Evaporator Outlet Temperature (°C)	17.616	0.071	0.029	0.071	0.150	0.166
	Compressor Power (W)	2913.00	5.37	2.19	5.36	7.28	9.04
	Liquid Line Temperature (°C)	37.764	0.013	0.005	0.013	0.150	0.151
	Liquid Line Pressure (kPa)	2700.09	5.85	2.39	5.85	3.50	6.81
Stage 2	Mass Flow Rate (g/s)	0.08161	0.00131	0.00054	0.00131	0.00082	0.00155
	Evaporator Outlet Pressure (kPa)	1159.90	0.99	0.41	0.99	3.50	3.64
	Evaporator Outlet Temperature (°C)	16.469	0.052	0.021	0.052	0.150	0.159
	Compressor Power (W)	3047.83	1.60	0.65	1.60	7.62	7.79
	Liquid Line Temperature (°C)	41.380	0.039	0.016	0.039	0.150	0.155
	Liquid Line Pressure (kPa)	2770.88	1.59	0.65	1.59	3.50	3.84
RTU	Fan Power (W)	991.33	33.23	13.57	33.20	2.48	33.29

The next step is to calculate the cooling capacity and COP from the measured values. Although the reported values listed in Table 6.1 were calculated using REFPROP 9, the difference between the upper and lower confidence limits were calculated using Engineering Equation Solver.   A short code, shown below, was written to calculate the propagation of uncertainty for the cooling capacity and COP, example shown uses the values listed in Table B.1

```
m_stage1=0.0815
T_liq1=40.657
P_liq1=2700.09
T_vap1=17.616
P_vap1=1171.17
W_comp1=2.913
h_liq1=ENTHALPY(R410A,T=T_liq1,P=P_liq1)
h_vap1=ENTHALPY(R410A,T=T_vap1,P=P_vap1)
Q1=m_stage1*(h_vap1-h_liq1)
COP1=Q1/W_comp1

m_stage2=0.0816
T_liq2=36.92
P_liq2=2770.88
T_vap2=16.469
P_vap2=1159.9
W_comp2=3.0478
h_liq2=ENTHALPY(R410A,T=T_liq2,P=P_liq2)
h_vap2=ENTHALPY(R410A,T=T_vap2,P=P_vap2)
Q2=m_stage2*(h_vap2-h_liq2)
COP2=Q2/W_comp2

W_fan=0.9913

Q_total=Q1+Q2
W_total=W_comp1+W_comp2+W_fan
COP_total=Q_total/W_total
```

Results:

Variable±Uncertainty **Partial derivative** **% of uncertainty**

Q1 = 13.32±0.171
m_{stage1} = 0.0815±0.00103 $\partial Q1/\partial m_{stage1}$ = 163.5   96.97 %
P_{liq1} = 2700±6.81 $\partial Q1/\partial P_{liq1}$ = 0.0001076   0.00 %
P_{vap1} = 1171±3.61 $\partial Q1/\partial P_{vap1}$ = -0.002226   0.22 %
T_{liq1} = 40.66±0.151 $\partial Q1/\partial T_{liq1}$ = -0.1573   1.93 %
T_{vap1} = 17.62±0.166 $\partial Q1/\partial T_{vap1}$ = 0.09659   0.88 %

COP1 = 4.574±0.0604
m_{stage1} = 0.0815±0.00103 $\partial COP1/\partial m_{stage1}$ = 56.13   91.61 %
P_{liq1} = 2700±6.81 $\partial COP1/\partial P_{liq1}$ = 0.00003692   0.00 %
P_{vap1} = 1171±3.61 $\partial COP1/\partial P_{vap1}$ = -0.000764   0.21 %
T_{liq1} = 40.66±0.151 $\partial COP1/\partial T_{liq1}$ = -0.05401   1.82 %
T_{vap1} = 17.62±0.166 $\partial COP1/\partial T_{vap1}$ = 0.03316   0.83 %
W_{comp1} = 2.913±0.00904 $\partial COP1/\partial W_{comp1}$ = -1.57   5.52 %

Q2 = 13.84±0.2644
m_{stage2} = 0.0816±0.00155 $\partial Q2/\partial m_{stage2}$ = 169.6   98.79 %
P_{liq2} = 2771±3.84 $\partial Q2/\partial P_{liq2}$ = 0.00007928   0.00 %
P_{vap2} = 1160±3.64 $\partial Q2/\partial P_{vap2}$ = -0.002275   0.10 %
T_{liq2} = 36.92±0.155 $\partial Q2/\partial T_{liq2}$ = -0.1499   0.77 %
T_{vap2} = 16.47±0.159 $\partial Q2/\partial T_{vap2}$ = 0.09728   0.34 %

COP2 = 4.539±0.08753
m_{stage2} = 0.0816±0.00155 $\partial COP2/\partial m_{stage2}$ = 55.63   97.05 %
P_{liq2} = 2771±3.84 $\partial COP2/\partial P_{liq2}$ = 0.00002601   0.00 %
P_{vap2} = 1160±3.64 $\partial COP2/\partial P_{vap2}$ = -0.0007464 0.10 %
T_{liq2} = 36.92±0.155 $\partial COP2/\partial T_{liq2}$ = -0.04919   0.76 %
T_{vap2} = 16.47±0.159 $\partial COP2/\partial T_{vap2}$ = 0.03192   0.34 %
W_{comp2} = 3.048±0.00779 $\partial COP2/\partial W_{comp2}$ = -1.489   1.76 %

Q_total = 27.16±0.3149
m_{stage1} = 0.0815±0.00103 $\partial Q_{total}/\partial m_{stage1}$ = 163.5   28.60 %
m_{stage2} = 0.0816±0.00155 $\partial Q_{total}/\partial m_{stage2}$ = 169.6   69.65 %
P_{liq1} = 2700±6.81 $\partial Q_{total}/\partial P_{liq1}$ = 0.0001076   0.00 %
P_{liq2} = 2771±3.84 $\partial Q_{total}/\partial P_{liq2}$ = 0.00007928   0.00 %
P_{vap1} = 1171±3.61 $\partial Q_{total}/\partial P_{vap1}$ = -0.002226   0.07 %
P_{vap2} = 1160±3.64 $\partial Q_{total}/\partial P_{vap2}$ = -0.002275   0.07 %
T_{liq1} = 40.66±0.151 $\partial Q_{total}/\partial T_{liq1}$ = -0.1573   0.57 %
T_{liq2} = 36.92±0.155 $\partial Q_{total}/\partial T_{liq2}$ = -0.1499   0.54 %
T_{vap1} = 17.62±0.166 $\partial Q_{total}/\partial T_{vap1}$ = 0.09659   0.26 %
T_{vap2} = 16.47±0.159 $\partial Q_{total}/\partial T_{vap2}$ = 0.09728   0.24 %

COP_total = 3.907±0.04946
m_{stage1} = 0.0815±0.00103 $\partial COP_{total}/\partial m_{stage1}$ = 23.52   23.98 %
m_{stage2} = 0.0816±0.00155 $\partial COP_{total}/\partial m_{stage2}$ = 24.39   58.41 %
P_{liq1} = 2700±6.81 $\partial COP_{total}/\partial P_{liq1}$ = 0.00001547   0.00 %
P_{liq2} = 2771±3.84 $\partial COP_{total}/\partial P_{liq2}$ = 0.0000114   0.00 %
P_{vap1} = 1171±3.61 $\partial COP_{total}/\partial P_{vap1}$ = -0.0003201 0.05 %
P_{vap2} = 1160±3.64 $\partial COP_{total}/\partial P_{vap2}$ = -0.0003272 0.06 %
T_{liq1} = 40.66±0.151 $\partial COP_{total}/\partial T_{liq1}$ = -0.02263   0.48 %
T_{liq2} = 36.92±0.155 $\partial COP_{total}/\partial T_{liq2}$ = -0.02157   0.46 %
T_{vap1} = 17.62±0.166 $\partial COP_{total}/\partial T_{vap1}$ = 0.01389   0.22 %
T_{vap2} = 16.47±0.159 $\partial COP_{total}/\partial T_{vap2}$ = 0.01399   0.20 %
W_{comp1} = 2.913±0.00904 $\partial COP_{total}/\partial W_{comp1}$ = -0.562   1.05 %
W_{comp2} = 3.048±0.00779 $\partial COP_{total}/\partial W_{comp2}$ = -0.562   0.78 %
W_{fan} = 0.9913±0.03329 $\partial COP_{total}/\partial W_{fan}$ = -0.562   14.30 %

www.ingramcontent.com/pod-product-compliance
Lightning Source LLC
Chambersburg PA
CBHW081606170526
45166CB00009B/2843